61

Advances in Biochemical Engineering/Biotechnology

Managing Editor: T. Scheper

Springer-Verlag Berlin Heidelberg GmbH

Biotechnology of Extremophiles

Volume Editor: G. Antranikian

With contributions by
G. Antranikian, M. Ciaramella, M. S. da Costa,
E. A. Galinski, R. Ladenstein, S. Maloney,
V. T. Marteinsson, M. Moracci, R. Müller,
J. van der Oost, F. M. Pisani, D. Prieur,
N. J. Russell, M. Rossi, H. Santos, R. Sharp,
W. M. de Vos

 Springer

This series represents critical reviews on the present and future trends in Biochemical Engineering/Biotechnology, including microbiology, genetics, biochemistry, chemistry, computer science and chemical engineering. It is adressed to all scientists at universities and in industry who wish to keep up-to-date in this extremly fast developing area of science.

In general, special volumes are dedicated to selected topics and are edited by well known guest editors. The managing editor and publisher will however always be pleased to receive suggestions and supplementary information. Manuscripts are accepted in English.

In references Advances in Biochemical Engineering/Biotechnology is abbreviated as Adv. Biochem. Engin./Biotechnol. as a journal.

ISSN 0724-6145

ISBN 978-3-662-14772-6   ISBN 978-3-540-69652-0 (eBook)
DOI 10.1007/978-3-540-69652-0

Library of Congress Catalog Card Number 72-152360

Typesetting: Fotosatz-Service Köhler OHG, Würzburg
Cover: Design & Production, Heidelberg
SPIN: 10573584   02/3020 – 5 4 3 2 1 0 – Printed on acid-free paper

Attention all "Enzyme Handbook" Users:

Information on this handbook can be found via the internet at

http://www.springer.de/chem/samsup/enzym-hb/ehb_home.html

At no charge you can download the complete volume indexes Volumes 1 through 13 from the Springer www server at the above mentioned URL. Just click on the volume you are interested in and receive the list of enzymes according to their EC-numbers.

Preface

The number of studies on extremophilic microorganisms has grown exponentially in the last few years. These exotic organisms (extremophiles) are adapted to living at 100 °C in volcanic springs, at low temperatures in the cold polar seas, at high pressure in the deep sea, at very low and high pH values (pH 0-1 or pH 10-11), or at very high salt concentrations (35 %). Recent developments clearly show that cell components of extremophilic archaea and bacteria are unique and deliver a valuable source of new biocatalysts and compounds. Since many industrial enzymes are required to function under extreme conditions, there is also a considerable commercial pressure to discover stable biocatalysts in modern biotechnology. Extremophiles and their cell components, therefore, are expected to play an important role in the chemical, food, pharmaceutical, paper and textile industries as well as environmental biotechnology.

In this special issue, review articles summarize the most outstanding features of hyperthermophilic, psychrophilic, barophilic and halophilic microorganisms. Further articles cover the protein structure and molecular biology of these exotic microorganisms. In addition, the potential applications of extremophiles are reviewed, including the production of enzymes, compatible solutes and the use of these extremophiles in the degradation of xenobiotics.

I am particularly indebted to the authors for their valuable contributions and to the publishers for their cooperation.

Hamburg, November 1997   Prof. Dr. G. Antranikian

Contents

Molecular Adaptations in Psychrophilic Bacteria: Potential for Biotechnological Applications

Nicholas J. Russell

Microbiology Laboratories, Department of Biological Sciences, Wye College University of London, Wye, Ashford, Kent TN25 5AH, England. *E-mail: n.russell@wye.ac.uk*

Bacteria which live in cold conditions are known as psychrophiles. Since so much of our planet is generally cold, i. e. below 5 °C, it is not surprising that they are very common amongst a wide variety of habitats. To enable them to survive and grow in cold environments, psychrophilic bacteria have evolved a complex range of adaptations to all of their cellular components, including their membranes, energy-generating systems, protein synthesis machinery, biodegradative enzymes and the components responsible for nutrient uptake. Whilst such a systems approach to the topic has its advantages, all of the changes can be described in terms of adaptive alterations in the proteins and lipids of the bacterial cell. The present review adopts the latter approach and, following a brief consideration of the definition of psychrophiles and description of their habitats, focusses on those adaptive changes in proteins and lipids, especially those which are either currently being explored for their biotechnological potential or might be so in the future. Such applications for proteins range from the use of cold-active enzymes in the detergent and food industries, in specific biotransformations and environmental bioremediations, to specialised uses in contact lens cleaning fluids and reducing the lactose content of milk; ice-nucleating proteins have potential uses in the manufacture of ice cream or artificial snow; for lipids, the uses include dietary supplements in the form of polyunsaturated fatty acids from some Antarctic marine psychrophiles.

Keywords: Psychrophilic Bacteria; biotechnology; cold-active enzymes; unsaturated fatty acids; low temperature.

Advances in Biochemical Engineering / Biotechnology, Vol. 61
Managing Editor: Th. Scheper
© Springer-Verlag Berlin Heidelberg 1998

1
Introduction

Many different types of organism have evolved the ability to survive or live at low temperatures, even below zero, using a wide range of strategies, but this review concentrates on the ways in which bacteria have adapted their protein and lipid structures to work efficiently in the cold. This approach is adopted, rather than the systems one [1], because ultimately all the metabolic and structural systems within a bacterial cell are dependent upon the structure and activity of proteins and lipids. (Carbohydrates might also be included, but since there is little or no evidence that structural alterations in these polymers have any role to play in cold adaptation they are not considered further.) Due to their small size, bacteria are unable to insulate themselves and even motile species cannot relocate sufficient distance to use avoidance strategies of adaptation. Therefore, it follows that their macromolecules must adapt their structures per se in order to retain function at low temperature, and it is these adaptations which form the focus of this review.

Changes in proteins of psychrophiles are genotypic characters, fixed in the genome and expressed as enzymes which have the ability to function at low temperatures. Although transcription and translation are regulated by temperature and the synthesis of some proteins (e. g. cold-shock proteins) may be specifically induced by a shift to low temperature, the phenomenon of cold-shock (and heat-shock) is common to psychrophiles, mesophiles and thermophiles alike. There is no evidence that the enzymes and structural proteins of psychrophiles alter their primary structure in a temperature-dependent manner. Some cellular enzymes will be those of lipid biosynthesis and temperature-dependent changes in their activity (and for some their expression also) will lead to phenotypic alterations in the lipid composition of psychrophiles when their growth temperature is lowered. Such phenotypic changes are superimposed on the genotypic determinants of lipid composition. Thus in relation to growth temperature there are both genotypic and phenotypic differences in lipid composition, but only genotypic differences in protein structure, between psychrophiles and mesophiles/thermophiles.

The topic of psychrophiles has been reviewed at regular intervals (e. g. see [2–10]) and is the intrinsic subject of two books on Antarctic microbiology [11, 12]. Over the years the emphasis has shifted from defining growth characteristics to exploring enzyme regulation by temperature, to the maintenance of membrane fluidity at low temperature, and most recently to the molecular basis of cold activity of enzymes. The latter might be considered as the "secret of psychrophily". However, only in recent years have molecular biological techniques been used to isolate the genes of psychrophilic proteins in order to produce sequence comparisons with homologous proteins from mesophiles and thermophiles [13 – 15]. Only very recently has the first enzyme from a psychrophilic bacterium been crystallised so as to facilitate a proper comparative structural investigation using X-ray diffraction [16]. A major driving force for the current interest in psychrophiles is the belated realisation of their immense biotechnological potential [8, 17].

This review concentrates on recent advances in our understanding of protein structure and the lipids in psychrophilic bacteria, which have potential for biotechnological exploitation as "cell factories" for the biosynthesis of commercially-useful products. It should be emphasised that such work on proteins and lipids only progresses within a framework of studies on other (equally important) work on the regulation of intracellular metabolism, gene expression and (for extracellular enzymes) protein secretion in psychrophiles. The importance of that work is reflected in the biotechnological significance of the development of psychrophilic expression systems. Discussion of these molecular aspects are put in context by a brief consideration of what defines a psychrophile and where they are found, and is followed by a review of the biotechnological applications of the organisms and their products.

2
Setting the Limits

2.1
Definition of Psychrophiles

Notwithstanding that all definitions are artificial and that Nature may ignore the rules we care to choose, it is nonetheless useful to provide a framework for ideas of what constitutes a cold-adapted bacterium, and indeed there does appear to be a class of microorganisms including all the major groupings (i.e. bacteria, yeasts, algae and fungi) which are capable of growing at or very close to zero (i.e. 0 °C). As the following will show, such psychrophiles do possess special adaptations of protein sequence and folding, and have features of their membrane lipid composition which enable them to function efficiently at temperatures inimical to mesophiles (and thermophiles).

Following more than a decade of debate about what constitutes a psychrophile, in 1975 Morita [4] proposed that, besides having the ability to grow at or below zero, they have an optimum growth temperature below 15 °C and a maximum temperature below 20 °C. He further distinguished psychrophiles from psychrotrophs (psychrotolerants) on the basis of their cardinal temperatures, in that the latter have a minimum growth temperature which is at or just above zero, and optimum and maximum growth temperatures above 20 °C [4]. For the purposes of this review, the two groups will generally not be differentiated, and the single term "psychrophile" will usually be used for convenience. The possible biochemical differences between true psychrophiles and psychrotolerants have been discussed recently elsewhere [18]. Feller et al. [19] have also highlighted the important point that growth rate (usually used for measuring cardinal temperatures) may not be as relevant as growth yield, and that some psychrophiles produce more biomass at low temperatures even though the growth rate may be little different to that achieved at temperatures within the upper half of the growth temperature range. This fact is particularly relevant to the biotechnological exploitation of psychrophiles, in that strains should be selected which have optimal biomass production at low temperatures (rather than necessarily fast growth rates). In addition, the yield of extracellular enzymes may also be greatest at lower temperatures (e.g. see [19, 20]), reflecting the thermal characteristics of the secretion process [21, 22]. Recent studies of the psychrotolerant *Pseudomonas fluorescens* indicate that cellular protein production is biphasic in relation to growth temperature with an optimum corresponding to the junction between the high and low temperature domains [23].

2.2
Habitats of Psychrophilic Bacteria

Humans tend to populate the temperate and warm places, yet most of our planet is cold, seldom reaching temperatures above 5°C because nearly three-quarters of the Earth is covered by (deep) oceans, where the temperature is permanently about 3°C [24]. Psychrophilic bacteria are major contributors to nutrient cycling in the deep-oceans [25]. The deep-sea is also a high pressure environment so many bacterial isolates are not only psychrophilic but also barophilic or barotolerant [26–29]. The deep-ocean floor is overlain by sediments up to a kilometre deep, and viable psychrophilic bacterial populations have been demonstrated in five Pacific Ocean sites at depths of up to 500 m into the sediment [29]. Unique bacterial types have been found there and their rate of decline in numbers with depth shows that they may be present at even greater depths [30]. The continent of Antarctica as well as the land masses in the Arctic provide permanently cold terrestrial environments as well as an aquatic niche in the associated ice which melts in the summer to provide a habitat for the development of vast microbial "blooms" [11, 31–33]. The soils and ice (glaciers) in Alpine regions also harbour psychrophilic microorganisms [34]. In addition to these obviously cold places, there are many more readily accessible and visible soils and waters which become cold both diurnally and seasonally, from which psychrophiles can be isolated using appropriate low temperature techniques.

Psychrophilic microorganisms, including bacteria, yeasts, fungi and microalgae, can be found in soils, in waters (fresh and saline, still and flowing) and associated with plants and cold-blooded animals. They present, therefore, a broad microbial diversity. Bacterial psychrophiles are more likely to be Gram-negative than Gram-positive, some of the genera which are frequently isolated being *Pseudomonas, Achromobacter, Flavobacterium, Psychrobacter* and *Cytophaga* [35]. However, although Franzmann et al. [36] have isolated methanogens from Antarctic lakes, and methanogenesis can be detected at low temperatures in cold Arctic tundra soils [37], the archaea appear to be poorly represented amongst psychrophilic populations. A recent report that archaea are not only abundant in picoplankton isolated from water and frazil-ice immediately beneath pack-ice in the Southern ocean but also contribute to its productivity [38] is misleading in this context because no archaeal taxa were cultivated and the conclusions were based on rRNA probe analysis which revealed relationships to known (hyper)thermophilic archaea. Nonetheless, the ability to grow at low temperature seems to have been acquired during evolution by all of the major groupings of microorganisms, and the range of species within a particular cold habitat will reflect those many different parameters (e.g. primary nutrient, ability to withstand desiccation, pH, salinity, etc.) to which an organism must adapt if it is to compete successfully with others.

Some psychrotolerant bacteria are pathogenic; they include strains of *Clostridium botulinum* and *Bacillus cereus*, which can grow and produce toxins in food stored in chill-cabinets, causing food-poisoning by intoxication upon subsequent ingestion of the contaminated product. *Bacillus cereus* as well as *Listeria monocytogenes* and *Yersinia enterocolytica* have unusually broad growth temperature ranges from near zero to above 40 °C; they are found in the natural environment and can contaminate food stored at low temperatures and then infect man, growing rapidly in the host and producing toxin [39]. In the plant world, a number of psychrotolerant pseudomonads, grouped together as *Pseudomonas syringae* strains, as well as *Erwinia* spp. are pathogenic. They cause a variety of diseases, including soft rots and wilts. As part of the pathogenic mechanism, the pseudomonads also produce ice-nucleation proteins, which have biotechnological potential (see Sect. 4.2).

The widespread use of refrigeration to store fresh and otherwise preserved foodstuffs provides a great diversity of nutrient-rich habitats for some well-known psychrotolerant food-spoilage bacteria, including *Brochothrix thermosphacta* and *Pseudomonas fragi* [40]. It is due to the metabolic activities of such bacteria, fungi and yeasts that refrigerated food has a limited storage time, despite the fact that it may also be preserved in some additional way (e.g. by the addition of salt or a modified gas atmosphere). For example, *B. thermosphacta* can grow not only at chill-cabinet temperatures but is also resistant to high carbon dioxide and low-oxygen conditions [41]. These principles are well illustrated by cook-chill foods, which have risen in popularity recently [42]. These products contain minimal amounts of preservatives, so they have a limited shelf-life even at refrigerated temperatures.

3
Molecular Adaptations

3.1
Protein Structure

In order for a psychrophile to grow well at low temperatures, it must contain enzymes which have a high specific activity in the cold. This is generally reflected in a relatively low apparent optimum temperature for activity compared with the corresponding enzymes from mesophiles/thermophiles (reviewed in [3–5]). These are usually 20–30 centigrade degrees lower than the corresponding enzyme from a mesophile, and for some enzymes the apparent optimum may be as low as 15–20 °C (e.g. see [43]). Hand-in-hand with that property is the fact that the enzymes of psychrophiles are also generally thermolabile. For example, the activity of an amylase isolated from a psychrophile found in sediment from the Japan Sea was destroyed completely after 1 h incubation at 45 °C, and some inactivation was observed as low as 25 °C [29]. In a 10 min incubation, the purified amylase had maximal activity at 35 °C, retaining about 40 and 10% of the maximal activity, at 10 and 0 °C, respectively. Another amylase from a different isolate was fully inactivated within 30 min at 50 °C, and inactivation was observed as low as 30 °C in a 30 min incubation. The enzyme had maximal activity at 25 °C and about 35% of the activity was observed at 10 °C and 15% of the optimal activity remained at 0 °C [29].

Although "optimum temperature" has no formal basis in enzyme kinetics, studies of this parameter and related thermal properties demonstrated that, whilst psychrophilic enzymes may be protected by binding to their substrates or cofactors in the same way as are those from other organisms, the ability to work at low temperatures was inherent in the protein structure and did not depend on stabilisation by other molecules or ions. Nor is there evidence that particularly slow turnover of proteins in psychrophiles is the reason for their ability to grow at low temperatures (just as fast turnover is not implicated in thermophily), although protein turnover may increase rapidly when the upper temperature limit is approached, as demonstrated in *Arthrobacter globiformis SI55* by Potier et al. [44]. Furthermore it was found that the relatively low upper growth temperature limit of psychrophiles was due to the thermolability of one or a few key enzymes which lost activity rapidly at the upper growth temperature limit (e. g. see [45]).

Thus it became clear that the basis of psychrophily resided in protein structure. However, the techniques of molecular biology were then less widely applicable, and no structural investigations using such determinative methods as X-ray diffraction were performed. Instead, attention turned to thermophilic proteins and it was comparisons between the proteins of thermophiles and mesophiles which contributed to current understanding of thermal adaptation in terms of the molecular structure and dynamics of enzymes [46, 47]. Comparable studies on enzymes from psychrophiles have lagged behind, but some structural analyses based on gene sequences are now available and for the first time an enzyme from a psychrophilic bacterium has been crystallised [16], thereby paving the way for direct structural observations. Several laboratories are now applying the same techniques of gene cloning and overexpression, followed by enzyme purification and crystallisation, to psychrophilic proteins as have been used widely for thermophilic proteins: thus, more structural comparisons should soon become available. What follows is based, therefore, on deductions about protein molecular architecture made from structures which have been predicted from gene nucleotide sequence data – these are available for lactate dehydrogenase, proteinases (subtilisins), lipase, α-amylases, triosephosphate isomerase, β-galactosidase, isocitrate dehydrogenase, β-lactamase, citrate synthase and esterase from psychrophilic bacteria (Table 1).

The first of such comparisons was published by Zuber and co-workers, who were also the first to sequence an enzyme from a psychrophilic bacterium, the lactate dehydrogenase (LDH) from *Bacillus psychrosaccharolyticus* [13]. Subsequently, the gene was cloned and sequenced [48]. The data were subjected to

Table 1. Enzymes from psychrophilic bacteria which have been cloned and sequenced

Enzyme	Bacterium	Reference
Lactate dehydrogenase	*Bacillus psychrosaccharolyticus*	48
Subtilisin (proteinase)	*Bacillus* sp. TA39	59
	Bacillus sp. TA41	20
Lipase[a]	*Moraxella* sp. TA144	50
	Psychrobacter immobilis	51
α-Amylase	*Alteromonas haloplanctis*	52
Triosephosphate isomerase	*Moraxella* sp. TA137	14
β-Galactosidase[b]	*Arthrobacter* sp. B7	53, 54
Isocitrate dehydrogenase[c]	*Vibrio* sp. ABE-1	55
β-Lactamase	*Psychrobacter immobilis*	19
Citrate synthase	*Moraxella* sp. DS2-3R	56
Esterase	*Pseudomonas* sp. LS107d2	57

[a] Three isozymes were cloned
[b] Several isozymes exist and one was cloned
[c] Two isozymes exist

specific matrix analysis in order to compare the LDH enzyme sequence from the psychrophilic *B. psychrosaccharolyticus* with those of the mesophilic *B. megaterium* and thermophilic *B. stearothermophilus*. The analysis demonstrated that the psychrophilic LDH has more polar and charged residues and less hydrophobic and ion-pairing residues, which should give the active site more flexibility but greater stability at low temperatures [47, 58]. It is important to emphasise the relationship between enzyme stability and flexibility, in relation to psychrophilic adaptations [19]. Hydrogen bonds and van der Waals and electrostatic interactions are formed exothermically, so they are strengthened by low temperatures. In contrast, hydrophobic bonds are formed endothermically and so will be weakened by low temperatures. Therefore, at low temperatures the weakening of hydrophobic bonds will threaten denaturation. In addition, kinetic considerations present a problem of retaining enzyme activity at low temperatures. Thus a balance must be struck between the need to stabilise the enzyme, yet retain sufficient conformational flexibility for enzyme activity at low temperatures. Subsequent to the pioneering work by Zuber, Gerday's group has made similar observations on a wider range of enzymes in a comparison of the gene sequences of psychrophilic α-amylase, proteinase, β-lactamase and lipase, with corresponding mesophilic sequences from bacterial and animal sources [15, 19, 59, 60]. The examples of α-amylase and proteinase (subtilisins) are considered now in more detail.

The α-amylase of a strain of *Alteromonas haloplanctis* isolated from Antarctic sea-water has an apparent optimum temperature for activity which is >30 centigrade degrees lower than that of a mesophilic (pig) α-amylase with which it shares 53% sequence identity (and 66% sequence similarity) [19, 61]. Several biophysical parameters of the psychrophilic enzyme point towards it having a looser conformation: for example, it exhibits faster denaturation with heating or the addition of urea or some chaotropic agents. Based on determinations of k_{cat} and K_m it has a catalytic efficiency which is seven-fold greater at 4 °C than the pig enzyme, yet it has the same three key amino acids (one glutamate and two aspartate residues) in the active site and has essentially the same overall three-domain architecture, typical of other amylases, comprising an $(\alpha/\beta)_8$ barrel, a β-pleated sheet domain and a globular domain of two facing β-sheets [61]. The more flexible, heat-labile conformation comes from large reductions (32–63%) in the numbers of hydrophobic and ionic/electrostatic interactions within the psychrophilic molecule – i.e. it is less tightly bonded compared with the meso-philic enzyme. These changes are summarised in Table 2. Most occur in the $(\alpha/\beta)_8$ barrel domain and, apart from the small reduction in aromatic-aromatic interactions, they will promote conformational flexibility. There is an overall sharp decrease in the core hydrophobicity of the molecule, whilst most of the substitutions of proline are by small residues such as alanine and occur within the loops connecting secondary structures, thereby increasing their flexibility. Although the active residues in the catalytic site are conserved, it may have greater flexibility since the substitution of of Arg[158] by Gln will weaken contacts between the separate domains around the catalytic cleft; this could be important in preserving substrate accessibility at lower temperatures when less kinetic energy is available.

The psychrophilic α-amylase has recently been crystallised and X-ray crystal-lographic studies initiated [16], which together with site-directed mutagenesis of the native enzyme will lead to further molecular insights of the basis of psychrophily in this enzyme.

Currently, in economic terms, the major industrial enzymes are the subtilisin-like proteinases; many have been purified, cloned and sequenced, and the three-dimensional structures of some are known [62]. Two cold-active subtilisin genes have been cloned from psychrotolerant *Bacillus* spp. isolated from Antarctic sea-water [20, 49]. The two enzymes each have 309 amino acids and display 92% sequence identity, and may be considered as isozymes; they are encoded by adjacent genes, present in both *Bacillus* spp., but one species expresses one iso-zyme and the other species the second isozyme [19].

Table 2. Psychrophilic adaptations in the α-amylase of *Alteromonas haloplanktis* A23 compared with pig (mesophilic) α-amylase

Parameter	Reduction (%) in psychrophilic enzyme
Salt bridges	63
Amino-aromatic interactions	55
Oxygen-aromatic interactions	32
Aromatic-aromatic interactions	5
Sulphur-aromatic interactions	60
Hydrophobic clusters	54
Proline content	38
Arginine content	55

Adapted from Feller et al. [61].

The size of the psychrophilic subtilisins is some 30–40 amino acids bigger than their mesophilic counterparts, but the overall three-dimensional structure (deduced from homology modelling of the gene sequence) is similar, with essentially the same secondary structure of nine α-helices and eight β-strands – it is in the interconnecting loops where the extra residues are found, up to 12 per loop, giving rise to a more flexible structure with the increased freedom of movement of the secondary structures.

The psychrophilic subtilisins contain a disulphide bridge which is absent from mesophilic subtilisins and might be expected to stabilise the protein. However, this is not so, because the thermostability (thermolability) of the psychrophilic isozymes is unchanged whether the disulphide bridge is intact or broken by reduction.

The homology models indicate that, compared with the mesophilic subtilisins, the psychrophilic enzymes have fewer salt bridges and aromatic-aromatic interactions, and more solvent interactions. The catalytic "triad" of aspartic acid, histidine and serine residues, giving the typical "charge-relay" catalytic mechanism of serine proteinases [63], is preserved in the cold-adapted enzyme. In contrast, there appear to be significant differences in the substrate-binding sites. In one loop, connecting a β-strand and an α-helix, a proline residue is substituted with a phenylalanine residue, giving more flexibility to the helix, which significantly includes the active site serine [59]. A second substitution of a bulky residue (tyrosine or valine) at position 104 in mesophilic subtilisins with the smaller alanine may also contribute to making substrate access to the active site easier in psychrophilic subtilisins.

The cold-active subtilisins have a much lower affinity for an essential Ca^{2+} which is due to the substitution of an asparagine residue (a good ligand for Ca^{2+}) with threonine (a poor ligand). In order to test whether this was indeed a psychrophilic adaptation, site-directed mutagenesis was used to replace the threonine with aspartic acid which resulted in stabilisation of the psychrophilic subtilisin and gave it better activity at low temperatures [19]. This experiment demonstrated that it is indeed feasible to use a molecular approach to engineer improved properties in psychrophilic enzymes for biotechnological applications.

Thus, in summary, a simplistic view might be that, since the requirements for flexibility at low temperatures of psychrophilic enzymes apparently are the opposite of the requirements for conformational stability at high temperatures needed in thermophilic proteins, the structural differences would be mirror images of each other in the two classes of protein. There is some evidence for a continuum of structural changes (e.g. for psychrophilic α-amylase compared with mesophilic counterparts [61]). However, a broader look at all the available gene sequences does not support that contention, and instead a wide spectrum of molecular changes which are not necessarily the opposite of those in thermophilic proteins is becoming apparent. For instance, analysis of the triose-phosphate isomerase genes from psychrophilic and thermophilic bacteria exemplifies the differences between their molecular adaptations: the thermophilic protein is stabilised largely by modifications of the α-helical regions through the helix dipole and helix-forming residues, whereas the adaptation to low temperature of the psychrophilic enzyme is through changes in the helix-capping residues [14]. Depending on how the native three-dimensional structure is achieved, different psychrophilic enzymes have evolved a range of molecular adaptations, including additional glycine residues, a low argini-ne/lysine ratio, more hydrophilic surfaces, the lack of several salt bridges and fewer aromatic interactions, in order to cope with the exigencies of low tempe-rature. As more psychrophilic enzymes and their genes are analysed, this diver-sity is likely to increase.

For (thermophilic) enzymes which contain more than one polypeptide chain, an important feature of stability is the extent and water-accessibility of hydro-phobic interactions between the subunits (e.g. see [64]). The counterpart of this aspect has been little studied in psychrophilic enzymes, but a good example is provided by the two isocitrate dehydrogenase (ICDH) isozymes in the psychro-philic *Vibrio* sp. strain ABE-1: the dimeric ICDH-I is relatively more thermo-stable than ICDH-II which is a monomer and therefore lacks the additional sta-bilising hydrophobic quaternary interactions [65]. Based on values of V_{max}/K_m as a measure of the catalytic efficiency, ICDH-II is approximately 70-fold more active than ICDH-I at low temperatures [66]. There is also evidence that the activity of ICDH-I may be regulated by phosphorylation [67], which is interesting in view of the report of temperature-dependent phosphorylation of three membrane proteins in an Antarctic isolate of *Pseudomonas syringae* by Ray et al. [68] and the known involvement of protein phosphorylation in cellu-lar regulation.

In addition to those of ICDH, isozymes have been detected in psychrophiles for lipase and β-galactosidase (Table 1); they may also exist for citrate synthase and lactate dehydrogenase. Two of the isozymes of β-galactosidase have quite distinct temperature optima, one at about 20 and the other at 45 – 50 °C [53, 54], thus raising the possibility of thermally-controlled gene expression as a means of modulating total enzyme activity across the growth temperature range. This is a feature of regulation of ICDH in *Vibrio* sp. ABE-1, since more ICDH-II is synthesised at low temperatures by selective expression of the *icdII* gene (see Sect. 4.4).

3.2
Membrane Lipids

Publication of the Fluid-Mosaic Model of membrane structure in 1972 [69] focussed attention on the requirement for a fluid membrane for normal cellular growth, and in relation to psychrophiles it strengthened ideas that the main-tenance of membrane fluidity and permeability have quintessential roles in adaptation to growth at low temperatures. Therefore, it is not surprising that temperature-dependent changes in membrane lipid fatty acyl composition have been documented extensively, particularly because this is such an easy para-meter to quantify using gas-liquid chromatographic analysis of derived fatty acid methyl esters. These changes, their biochemical basis, and the relationships to membrane fluidity and permeability have been reviewed elsewhere [18, 70, 71]. There is no fatty acid composition which is unique to psychrophilic bacteria, and different taxa have evolved quite distinct strategies for maintaining a membrane that is fluid enough at low temperatures whilst retaining sufficient molecular order to prevent excessive increases in passive permeability. Studies have generally concentrated on how the provision of fluidity is achieved; the preservation of adequate molecular order has largely been ignored. However, it is this latter aspect which may be more significant from a biotechnological perspective, and some of the mechanisms for achieving lipid order at low temperatures are highlighted below.

A decrease in bacterial growth temperature leads to an increase in fatty acid unsaturation, or a decrease in average chain length, or an increase in methyl branching, or an increase in the ratio of *anteiso*-branching relative to *iso*-branching, or to some combination of these changes [7, 70]. The commonest alteration is in the amount of unsaturation and this also gives quantitatively the largest shift in the gel-to-liquid-crystalline phase transition temperature. The type of changes which occur will also reflect the taxonomic status of the organism, since methyl-branched fatty acids are much more common in Gram-positive bacteria compared with Gram-negatives [72, 73]. Lipids containing *anteiso*- or *iso*-branched fatty acyl chains have an "intermediate fluidity" with a relatively more ordered liquid-crystalline phase but a relatively more disordered gel phase, the effect of *anteiso*-methyl branches being greater than that of *iso*-methyl branches [74, 75]. The methyl branches are introduced during de novo fatty acid biosynthesis by the use of branched-chain primers [72]. Therefore, to invest this property in another organism by molecular genetic means would require the transfer of all the fatty acid synthetase genes; even if this could be achieved, it would still be necessary to marry the associated metabolic activities with those of the main synthetase. A better approach would be to transfer a fatty acid methylation gene, if this could be identified and cloned; there are some examples of bacteria which contain fatty acids methylated at other positions (i. e. besides *iso*- or *anteiso*-methyl branches) [73], but little is known about the methylation mechanism or if a single gene product is involved [72].

The majority of studies of thermal effects on bacterial fatty acids have measured the overall composition, ignoring the isomeric distribution, a structural feature which is seldom considered despite the fact that switching of the acyl chains between the *sn*-1 and *sn*-2 positions of phospholipids can alter their gel-to-liquid-crystalline phase transition temperature by up to 7 centigrade degrees [71]. The physiological significance of this value is that it represents 25–35% of the growth temperature span of a microorganism. The psychrotolerant bacterium *Psychrobacter uratovorans* (formerly known as *Micrococcus cryophilus*) has a *sn*-1/*sn*-2 distribution of fatty acyl chains which gives the lower melting point phospholipid isomers, and this distribution is maintained throughout the growth temperature range even after a sudden drop in temperature [76]. This arrangement, together with the highly unsaturated fatty acid composition of the organism, gives a gel-to-liquid-crystalline phase transition temperature that is well below zero whilst preventing the formation of deleteriously too strong lipid-lipid interactions, particularly at the lower end of the growth temperature range when there is a large proportion of dipalmitoleoyl phospholipid.

Bacteria do not usually contain polyunsaturated fatty acids, so temperature-dependent changes in the unsaturation of their lipids are generally in the proportions of monounsaturated components [72]. However, a notable exception to this general rule is a group of marine psychrophilic bacteria which contain significant proportions (up to 20%) of 20:5ω3, 22:6ω3 and 20:4ω6 (e.g. see [77–79]). Moreover, these polyunsaturated fatty acids (PUFA) have double-bond configurations which are typical of the ω3 and ω6 fatty PUFA typical of animals, including humans, a fact which makes them of considerable biotechnological interest and could lead to their exploitation as dietary supplements (see Sect. 4.5). The biological explanation of the occurrence of PUFA in these marine psychrophiles may be to balance the requirement for a liquid-crystalline ("fluid") membrane at low temperatures with the retention of a requisite level of order, as described by the "Dynamic Phase Behaviour Model" which is discussed by Nichols et al. [35]. Significantly, in terms of biotechnological manipulation, it has been shown in some marine psychrophiles that the proportion of PUFA increases in response to a decrease in growth temperature: for example, *Shewanella gelidimarina* isolated from Antarctic sea-ice responds to a decrease in growth temperature by not only decreasing the proportion of odd-chain fatty acids and increasing the even-chain fatty acids but also by increasing its content of 20:5ω3 to nearly 20% [80]. The PUFA content is also governed by the nature of the carbon source [81], which could be a means of manipulating their production.

Some bacteria are able to synthesise *trans*-unsaturated fatty acids, which offer protection against stresses such as heavy metals and toxic organic compounds [82]. Whilst uncommon generally amongst bacteria, this ability is a relatively common feature of marine psychrophiles, particularly those of the genus *Vibrio*. For example, the *Vibrio* sp. strain ABE-1 contains *trans* 16:1ω7, the proportion of which decreases with a lowering of the growth temperature [83]. This is consistent with the fact that lipids containing *trans*-unsaturated fatty acyl groups have physical properties which more closely resemble those of saturated rather than (*cis*) unsaturated counterparts [71]. Thus, in *Vibrio* sp. strain ABE-1 the *trans* 16:1ω7 effectively behaves as a saturated acid in order to balance the fluidising effects of *cis*-unsaturated fatty acyl chains. Another *Vibrio* sp., isolated from the gut of the Arctic char, contains both *cis* and *trans* 16:1ω7 as well as 20:5ω3 and a lowering of growth temperature causes an increase in the proportions of *trans* 16:1ω7 and 20:5ω3, but no change in *cis* 16:1ω7 [84]. It is assumed that the balance between *trans* 16:1ω7 and 20:5ω3 in response to thermal changes has the same effect, therefore, in maintaining the lipids in a liquid-crystalline phase, whilst at the same time balancing the dual requirements of "fluidity" and "order".

The mechanism of biosynthesis of *trans*-unsaturated fatty acid is by direct isomerisation of the *cis*-double bond in the acyl chain of a membrane lipid, without change of position [82]. Little is known about the biochemical mechanism, or whether a single gene product is involved. But in view of the ability of *trans*-unsaturated fatty acids to impart to bacteria resistance to a wide range of harsh conditions, transfer of the "*trans*-synthetase" gene from a marine psychrophile to biotechnologically-useful bacteria (or other organisms) would enhance their usefulness.

Changes in membrane lipid unsaturation in bacteria are mediated either by the so-called anaerobic pathway or by desaturase enzymes [72]. In the anaerobic pathway, a mixture of saturated and unsaturated fatty acids are produced by a modified set of fatty acid synthetase enzymes, i.e. by a process of de novo synthesis from acetyl-CoA, and the balance of saturated and unsaturated products is thermally regulated [85, 86]. In contrast, desaturases introduce a double bond into preformed fatty acids (usually as acyl chains in intact membrane lipids) by an aerobic process, which is also thermally regulated. Desaturases are widely distributed in most organisms, not only bacteria. They have proved difficult to investigate experimentally because they are membrane-bound enzymes and their function is mediated by several linked components which channel the hydrogen atoms removed from the substrate to molecular oxygen so as to form water in a manner analogous to that which occurs during respiration. This multi-component, membrane-bound structure makes them difficult to purify, and only recently has significant progress been made in the cloning of desaturase genes. For example, Murata's group cloned the $\Delta12$-desaturase *desA* gene from the chilling-resistant *Synechocystis* PCC6803 sp. and used it to transform the normally chilling-sensitive *Anacystis nidulans*, thus enabling it to synthesise diunsaturated fatty acids (mainly 16:1Δ9, 12) and making the organism and its photosynthetic machinery tolerant to low temperatures, thereby proving the link between cyanobacterial $\Delta12$-desaturase activity and the ability to grow at low temperatures [87, 88].

Theoretically, it should be possible to use molecular-genetic techniques to introduce cloned desaturase genes into industrially-useful microorganisms in order to give them a new capacity to synthesise, for example, polyunsaturated fatty acids or new monounsaturated fatty acids, so as to modify their cold-tolerance for biotechnological exploitation in low-temperature processes and reactions. This approach will need to be accompanied by a better understanding of the mechanism of fatty acid modification reactions and more insight into their (thermal) regulation. These aspects have been reviewed elsewhere [9].

4
Biotechnological Applications

Only within the past few years has it been recognised that psychrophilic microorganisms and their products or enzymes provide a large reservoir of potentially novel biotechnological exploitation [8, 10, 17, 89]. These applications and some of the benefits of psychrophiles and their products (enzymes) are summarised in Tables 3 and 4, and the following discussion highlights some in more detail.

4.1
Psychrophiles – Natural Cycles and Bioremediation

The widespread distribution of psychrophiles within natural and manmade environments reflects their broad metabolic capabilities: accordingly, psychrophiles may play any of the roles in the environment which have been attributed

Table 3. Some potential uses for psychrophiles and their products

Use	Psychrophilic source/product
Cold-water washing	Proteinase, lipase, cellulase
Meat tenderising	Proteinase
Food-processing	Carbohydrases
Flavour modification	Various specialised enzymes
Lactose hydrolysis in milk	β-Galactosidase
Food additives, dietary supplements	Polyunsaturated fatty acids
Environmental/on-line biosensors	Various enzymnes (e.g. oxidases)
Bioremediations	Various enzymes (e.g. oxidases)
Biotransformations	Various enzymes (e.g. dehydrogenases)
Contact-lens cleaning	Proteinase
5'-End-labelling of nucleic acids	Alkaline phosphatase

Table 4. Some advantages of psychrophiles and/or their enzymes in biotechnological applications

Rapid and economic process termination by mild heat treatment
Higher yields from reactions involving thermosensitive components
Modulation of the (stereo)specificity of enzyme-catalysed reactions
Cost-saving by elimination of expensive heating/cooling process-steps
Capacity for on-line monitoring under environmental conditions

to other microorganisms. There are no functions which are specifically reserved for the metabolic activities of psychrophiles, and they are major contributors to the primary cycles of nature, i.e. those for carbon, nitrogen, phosphate and sulphur [90]. For example, psychrophilic nitrogen-fixing rhizobia are known and some psychrophiles produce extracellular enzymes for the breakdown of macromolecules such as proteins and carbohydrates (e.g. chitin and cellulose) at low temperatures [91]. Reichardt observed that temperature optima for enzyme activities were higher than those for microbial growth [91], so enhanced rates of biosynthesis and secretion must be key features of (polymeric) organic matter cycling in cold environments (cf. Sect. 2.1). Low molecular weight environmental pollutants are also likely to be substrates for psychrophiles, which could be used for specific clean-up operations. For example, low temperature biodegradation of polyols (including ethylene, propylene and diethylene glycols) derived from aircraft de-icing fluids has been demonstrated in contaminated soil adjacent to an airport runway [92].

Thus, psychrophiles have the potential for use in engineered bioremediation systems operating in the open in places where the temperatures are permanently or cyclically (e.g. diurnally) low, specially in situations where the length of time for the process does not have to be short. Psychrophilic anaerobic digestion of human waste at 15 °C has been evaluated to pilot-plant scale, using bacterial seed cultures which had been adapted to low-temperature growth [93]. Although the process was slower than mesophilic/thermophilic digestions, taking 15 months for a suitably adapted psychrophilic population to develop, it was deemed satisfactory in terms of methane gas production and combustibility, and pH of the digested slurry. A full-scale low-temperature lagoon digester was found to operate successfully, giving 79% of the yield of biogas compared to a mesophilic digester of similar design [94]. Low-temperature digestions with an Antarctic *Arthrobacter* sp. have also been used to inactivate pathogens during aerobic and anaerobic treatments at 10 °C [95]. The efficient removal of nitrate has also been demonstrated in a psychrophilic anaerobic digester [96].

4.2
Proteins

A number of species of *Pseudomonas*, *Erwinia* and *Xanthomonas* are plant pathogens, causing damage to leaves and flowers by triggering ice crystal formation through the action of ice-nucleating proteins at subzero temperatures (i.e. –2 to –5 °C) when the water would otherwise remain supercooled and liquid [97]. The ice-nucleating proteins are all encoded by the *ina* genes and have three domains: a unique hydrophobic N-terminal domain; a unique hydrophilic C-terminal domain; and a central domain containing repeated octapeptide sequences [98]. A number of models have been proposed to explain how the proteins align water molecules to initiate ice formation [99, 100].

Several biotechnological uses are being explored for bacterial ice-nucleating proteins. These include ice-cream manufacture, making synthetic snow, and a range of food applications such as freeze-texturing, freeze-drying and concentrating [97]. An interesting variation is the use of "ice-minus" bacteria, which occur naturally or are genetically engineered. These mutants lack the *ina* genes, and controlled field experiments have demonstrated the potential for their use as frost protectants for sensitive plants by outcompeting the natural ice-nucleating pathovars [101].

4.3
Enzymes

Many of the naturally-occurring yeasts or bacteria on fruits and vegetables are psychrotolerant and perform some of the best known fermentations to give products like beer or cheese and other dairy foods, using production methods which are operated at "cool" temperatures. Strains have been selected by conventional means for improved cold-tolerance, because in order to engineer "psychrophily" genetically in an otherwise cold-sensitive organism would require the transfer of so many genes that it is not feasible at present. A recent example of a novel requirement for cold-tolerant yeasts is in the baking industry where the increasing use of frozen doughs makes it advantageous to use baking-yeast strains which retain their fermentative activity after freezing and storage in the cold [102].

There is a wide range of potential uses for cold-active enzymes, ranging from proteins, lipases and cellulases in cold-water washing programmes, to many applications in the food industry (Table 3). The engineering of psychrophilic subtilisins, which have huge commercial potential in the washing and cleansing industries, has already been proved to be possible, as discussed in Sect. 3.1.

Psychrophilic β-galactosidase could be used to lower the lactose content of milk (for lactose-intolerant persons) whilst it is held in cold storage; the current practice is to use a mesophilic enzyme which necessitates warming the milk which not only leads to organoleptic changes but also adds to the cost [53].

An inherent problem with biosensors (enzyme electrodes) is their instability under operational conditions, and the approach which has generally been used to overcome this drawback is either to immobilise the enzyme and/or to use a thermostable enzyme. However, it is not always convenient to increase the operating temperature and an alternative approach is to lower the temperature and use a psychrophilic enzyme. This has particular advantages in the on-line monitoring of environmental processes (e.g. wastewater treatment), pharmaceutical or food products stored in cold rooms, and in other low-temperature processes [103, 104].

4.4
Protein Expression Systems

In order to develop properly the biotechnological potential of psychrophiles, it is essential that psychrophilic expression systems are developed, so that cold-active proteins can be produced in vivo within an intracellular environment which is thermally appropriate for their folding pathway(s). Those genes for psychrophilic proteins which have been cloned to date have been expressed in *E. coli* cultured at 15–20°C, when growth is rather slow. Another good reason for developing psychrophilic expression systems is that they could well be used to alleviate the problem of inclusion body formation which plagues the expression of many proteins in mesophilic hosts; this can sometimes be lessened by lowering the growth temperature of *E. coli* or other mesophilic host – therefore, using a psychrophilic host at even lower temperatures should help further.

Although an homologous psychrophilic expression system has not yet been developed, there is information about transcriptional regulation and the translation apparatus in psychrophiles which is relevant and is therefore summarised below.

Bacteria (and other organisms) generally make proteins at rates which match their growth, so the ribosomal proteins of psychrophiles must be specially adapted to function at low temperature. This was demonstrrated in 1967 by Krajewska and Szer [105] who mixed ribosomal subunits from psychrophilic and mesophilic bacteria, and showed that psychrophilic ribosomes have low miscoding rates compared with those from mesophiles at the same temperature. Later work demonstrated that it is the initiation of protein synthesis which is particularly susceptible to inhibition by a sudden fall in temperature (see [7] for a summary). However, psychrophilic ribosomes and their component proteins were not subsequently subjected to a rigorous investigation of their structure.

Instead, insight into the cold regulation of protein synthesis has come from investigation of the phenomenon of cold-shock in both *E. coli* and some psychrophiles. After cold-shock there is suppression of house-keeping gene expression and the synthesis of a family of cold-shock genes [106]. The major cold-shock protein of *E. coli*, CspA, appears to be a transcriptional regulator of other cold-shock genes on the basis of its physiological effects [107] and structure [108, 109]. Homologous proteins have been identified in other bacteria, particularly in *B. subtilis* [110, 111], as well as in Antarctic psychrotrophic bacteria [112]. Cold-shock also leads to a decrease in the levels of the guanosine nucleotides (p)ppGpp inside the cell, the decrease being proportional to the magnitude of the downshift and the extent of the cold-shock response [113]. Since these molecules are made when ribosomes "idle", this observation was consistent with a model in which disruption of translation was a signal for the cold-shock response. It was supported by the finding that the RbfA protein, which is believed to interact with the 5'-helix of 16S rRNA and to function as a ribosomal maturation and/or translation initiation factor [114], was also a cold-shock protein [115]. It is postulated in the "cold-shock ribosome adaptation model" [115] that a sudden temperature fall blocks the initiation of translation of cellular mRNAs apart from those encoding cold-shock proteins, some of which (e.g. RbfA and CsdA) associate with the ribosomal subunits and 70S monomers to give polysomes which are able to translate efficiently at the new lower temperature.

However, such translational regulation of cold-shock cannot be the complete explanation, because the question remains as to the mechanism of continued production of cold-shock proteins after cold-shock. The CspA protein appears to regulate transcription by binding to ATTGG sequences in the promoters of cold-shock genes, and it has a high sequence similarity with the eukaryotic Y-box binding proteins which also regulate transcription [109]. This motif is commonly found in genes of *E. coli* which are inducible by low temperature [116]. It is present two base-pairs upstream from the -35 promoter of the *icdII* gene encoding the cold-active isocitrate dehydrogenase (ICDH) of *Vibrio* sp. ABE-1 ([55] and Sect. 3.1). This gene is selectively expressed at low temperatures, giving increased production of its mRNA [67], and confirmation of the role of the ATTGG sequence in low temperature inducibility comes from mutagenic experiments in which substitution of a single base results in complete loss of this property (N. Fukunaga, personal communication). Thus it appears that cold-shock genes are regulated by transcription, but no cold-specific sigma factor (analogous to σ^{32} in heat-shock) has been identified.

When bacteria are subjected to permanently cold conditions, a different set of proteins is synthesised, known as cold-acclimation proteins: these are synthesised continuously in response to isothermal growth at low temperature and are generally distinct from cold-shock proteins (e.g. see [117, 118]). Like the cold-shock response, the synthesis of cold-acclimation proteins is graded, with different sub-sets of proteins being made in response to different extents of low-temperature growth. The function of cold-acclimation proteins is unknown, but they are likely to be key components of an efficient low-temperature expression system in view of the sensitivity of the protein translation system to inhibition by cold.

4.5
Lipids

At the present time there are no examples of the biotechnological exploitation of bacterial lipids, but the marine psychrophiles (discussed in Sect. 3.2) could be a source of PUFA as dietary supplements. Humans require essential polyunsaturated fatty acids (PUFA), specifically linoleic acid (18:2 Δ9,12 or 18:2 ω6) and α-linolenic acid (18:3 Δ9,12,15 or 18:3 ω3), which are the precursors of longer-chain ω6 and ω3 PUFAs respectively such as arachidonic acid (AA, 20:4ω6) and docosahexaenoic acid (DHA, 22:6ω3) and eicosapentaenoic acid (EPA, 22:6 ω3). These are required for healthy growth either as precursors of such regulatory compounds as prostaglandins, thromboxanes and leukotrienes or as components of membrane acyl-lipids in brain and retina. Linoleic and α-linolenic acids are obtained from green vegetables and from plant oils, whereas fish oils are a major source of higher PUFA such as DHA. Recent evidence that the intake of these longer-chain PUFA helps to prevent diseases such as atherosclerosis, together with the world-wide threat to fish stocks, have combined to give a raison d'être and an impetus to the search for alternative (microbial) sources. The ease of culture of bacteria makes them attractive candidates, but the levels of PUFA (compared with those in fungi) will need to be increased to make them economically feasible.

5
Final Comments

The first psychrophilic bacteria were isolated by Forster in 1887 from preserved fish [119]. Even earlier, during the first Antarctic maritime explorations in the 1840s, the association of microalgae with sea-ice was noted from the colouration which they produce [120]. Despite these early discoveries, psychrophiles were little studied until the 1950s and 1960s, when investigations of the temperature dependence of growth and enzyme activity were performed. These studies were not followed up by structural investigations of crystallised enzymes, and even with the advent of molecular biological techniques for gene cloning, the investigation of psychrophilic proteins has lagged behind that of thermophiles.

However, in the European Commission funded project COLDZYME, there is now a coordinated effort of research in Europe into the molecular basis of cold-activity of proteins, and the physiology of psychrophiles as efficient producers of such enzymes for their biotechnological exploitation. Moreover, COLDZYME is part of a wider European initiative to develop a broad range of extremophiles and their products as biotechnological tools. So we may expect rapid advances in this field. For psychrophiles in particular, the first crystallisation of a cold-active enzyme, as described in Sect. 3.1, will pave the way for that progress.

6
References

1. Hochachka PW, Somero GN (1984) Biochemical adaptations. Princeton University Press, Princeton
2. Ingraham JL (1958) J Bacteriol 76:75
3. Ingram M (1965) Ann Inst Pasteur Paris 16:111
4. Morita RY (1975) Bacteriol Rev 29:144
5. Innis WE (1975) Annu Rev Microbiol 29:445
6. Herbert RA (1986) The ecology and physiology of psychrophilic micro-organisms. In: Herbert RA, Codd GA (eds) Microbes in extreme environments. Academic Press (for Society of General Microbiology, UK), London, p 1
7. Russell NJ (1990) Phil Trans Roy Soc London Series B 329:595
8. Russell NJ (1992) Physiology and molecular biology of psychrophilic micro-organisms. In Herbert RA, Sharp RJ (eds) Molecular biology and biotechnology of extremophiles. Blackie, Glasgow & London, p 203
9. Russell NJ (1997) Comp Biochem Physiol: In press
10. Gounot A-M (1991) J Appl Bacteriol 71:386
11. Vincent WF (1988) Microbial systems of Antarctica. Cambridge University Press, Cambridge
12. Friedmann EI (ed) (1993) Antarctic microbiology. Wiley-Liss, New York
13. Schlatter D, Kriech O, Suter F, Zuber H (1987) Biol Chem Hoppe-Seyler 368:1435
14. Rentier-Delrue F, Mande SC, Moyens S, Terpstra P, Mainfroid V, Goraj K, Lion M, Hol WGJ, Martial JA (1993) J Mol Biol 229:85
15. Arpigny JL, Feller G, Davail S, Narinx E, Zekhnini Z, Gerday C (1994) Adv Comp Environ Physiol 20:270
16. Aghajari N, Feller G, Gerday C, Haser R (1996) Prot Sci (1996) 5:2128
17. Margesin R, Schinner F (1994) J Biotechnol 33:1
18. Russell NJ (1993) Biochemical differences between psychrophilic and psychrotolerant microorganisms. In:Guerrero R, Pedros-Alio C (eds) Trends in microbial ecology. Spanish Society for Microbiology, Madrid, p 29
19. Feller G, Narinx E. Arpigny JL, Aittaleb M, Baise E, Genicot S, Gerday C (1996) FEMS Microbiol Lett 18:189
20. Davail S, Feller G, Narinx E, Gerday C (1992) Gene 119:143
21. Gugi B, Orange N, Hellio F, Burini JR, Hukllou C, Leriche F, Guespin-Michel JF (1991) J Bacteriol 173:3814
22. Guillou C, Merieau A, Trebert B, Guespin-Michel JF (1995) Biotechnol Lett 17:377
23. Guillou C, Guespin-Michel JR (1996) Appl Environ Microbiol 62:3319
24. Austin B (1988) Marine microbiology. Cambridge University Press, Cambridge
25. Morita RY (1980) Can J Microbiool 26:1375
26. Yayanos AA, Diez AS, Van Boxtel R (1981) Proc Natl Acad Sci USA 78:5212
27. Yayanos AA, DeLong EF (1987) Deep-sea bacterial fitness to environmental temperatures and pressures. In:Jannasch HW, Marquis RE, Zimmerman AM (eds) Current perspectives in high pressure biology. Academic Press, London, p 17
28. Jannasch HW, Wirsen CO, Taylor CD (1982) Science 216:1315
29. Hamamoto T, Russell NJ (1998) Psychrophiles. In:Horikoshi K, Grant WD (eds) Extremophiles. Wiley, New York:In press
30. Parkes RJ, Cragg BA, Bale SJ, Getliff JM, Goodman K, Rochelle PA, Fry JC, Weightman AJ, Harvey SM (1994) Nature 371:410
31. Wynn-Williams DW (1990) Adv Microbial Ecol 11:71
32. Vishniac HS (1993) The microbiology of Antarctic soils. In: Friedman EI (ed) Antarctic microbiology. Wiley-Liss, New York, p 297
33. Palmissano AC, Garrison DL (1993) Microorganisms in Antarctic sea ice. In:Friedman EI (ed) Antarctic microbiology. Wiley-Liss, New York, p 167
34. Gounot A-M (1976) Can J Microbiol 22:839

35. Nichols DS, Nichols PD, McMeekin TA (1995) Sci Progr Oxf 78:311
36. Franzmann PD, Springer N, Ludwig W, De Macario EC, Rohdes M (1992) System Appl Microbiol 15:573
37. Kotsyurbenko OR, Nozhevnikova AN, Soloviova TI, Zavarzin GA (1996) Ant van Leeuw Int J Gen Molec Microbiol 69:75
38. DeLong EF, Wu KY, Prézelin BB, Jovine RVM (1994) Nature 371:695
39. Walker SJ, Stringer MF (1990) Microbiology of chilled foods. In:Gormley TR (ed) Chilled foods, the state of the art. Elsevier, London, p 269
40. Russell NJ, Gould GW (1991) Factors affecting growth and survival. In:Russell NJ, Gould GW (eds) Food preservatives. Blackie, Glasgow, Chap 2
41. Gould GW, Russell NJ (1991) Major food-poisoning and food-spoilage micro-organisms. In:Russell NJ, Gould GW (eds) Food preservatives. Blackie, Glasgow, Chap 1
42. Bognar A, Bohling H, Fort H (1990) Nutrient retention in chilled foods. In:Gormley TR (ed) Chilled foods, the state of the art. Elsevier, London, p 305
43. Mitchell P, Yen HC, Mathemeier PF (1985) Appl Environ Microbiol 49:1332
44. Potier P, Drevet P, Gounot A-M, Hipkiss AR (1990) J Gen Microbiol 136:283
45. Malcolm NL (1969) Nature 221:1031
46. Zuber H (1988) Biophys Chem 29:171
47. Jaenicke R (1991) Eur J Biochem 202:715
48. Vckovski V, Schlatter D, Zuber H (1990) Biol Chem Hoppe-Seyler 371:103
49. Narinx E, Davail S, Feller G, Gerday C (1992) Biochim Biophys Acta 1131:111
50. Feller G, Thiry M, Gerday C (1991) DNA Cell Biol 10:381
51. Arpigny JL, Feller G, Gerday C (1993) Biochim Biophys Acta 1171:331
52. Feller G, Lonhienne T, Deroanne C, Libiouille C, Van Beeumen J, Gerday C (1992) J Biol Chem 267:5217
53. Trimbur DE, Gutshall KR, Prema P, Brenchley JE (1994) Appl Environ Microbiol 60:4544
54. Gutshall KR, Trimbur DE, Kasmir JJ, Brenchley JE (1995) J Bacteriol 177:1981
55. Ishii A, Susuki M, Sahara T, Takada Y, Sasaki S, Fukunaga N (1993) J Bacteriol 175:6873
56. Gerike U, Russell NJ, Hough D, Danson MJ (1997) Eur J Biochem 248:49
57. McKay DB, Jennings MP, Godfrey EA, MacRae IC, Rogers PJ, Beacham IR (1992) J Gen Microbiol 138:701
58. Zuber (1990) p 610 of Discussion in Russell (1990)
59. Davail S, Feller G, Narinx E, Gerday C (1994) J Biol Chem 269:17–448
60. Feller G, Zekhnini Z, Lamotte-Brasseur J, Gerday C (1997) Eur J Biochem: In press
61. Feller G, Pazan F, Theys F, Qian M, Haser R, Gerday C (1994) Eur J Biochem 222:441
62. Siezen RJ, De Vos WM, Leunissen JAM, Dijkstra BW (1991) Prot Eng 4:719
63. Stryer L (1981) Biochemistry, 2nd edn. Freeman, San Francisco, Chap 8
64. Wigley DB, Clarke AR, Dunn CR, Barstow DA, Atkinson T, Chia WN, Muirhead H, Holbrook JJ (1987) Biochim Biophys Acta 916:145
65. Ochiai T, Fukunaga N, Sasaki S (1979) J Biochem 86:377
66. Ochiai T, Fukunaga N, Sasaka S (1984) J Gen Appl Microbiol 30:479
67. Susuki M, Sahara T, Tsuruha J-I, Takada Y, Fukunaga N (1995) J Bacteriol 177:2138
68. Ray MK, Kumar GS, Shivaji S (1994) Microbiology 140:3217
69. Singer SJ, Nicolson GL (1972) Science 175:720
70. Russell NJ (1984) Trends Biochem Sci 9:108
71. Russell NJ (1989):Functions of lipids:structural roles and membrane functions. In: Ratledge C, Wilkinson SG (eds) Microbial lipids, vol 2. Academic Press, London, p 279
72. Harwood JL, Russell NJ (1984) Lipids in plants and microbes. George Allen & Unwin, London
73. Ratledge C, Wilkinson SG (1988) Microbial lipids, vol. 1. Academic Press, London
74. Macdonald PM, McDonough B, Sykes BD, McElhaney RN (1983) Biochem 22:5103
75. Macdonald PM, Sykes BD, McElhaney RN (1985) Biochem 24:2412
76. McGibbon L, Russell NJ (1983) Curr Microbiol 9:241
77. Ring E, Sinclair PD, Birkbeck H, Barbour A (1992) Appl Environ Microbiol 58:3777
78. Nichols DS, Nichols PD, McMeekin TA (1993) Antarct Sci 5:149

79. Hamamoto T, Takata N, Kudo T, Horikoshi K (1994) FEMS Microbiol Lett 119:77
80. Nichols DS, Nichols PD, Russell NJ, McMeekin TA (1997) Biochim Biophys Acta 1347:164
81. Nichols DS, Russell NJ (1996) Microbiology 142:747
82. Keweloh H, Heipieper HJ (1996) Lipids 31:129
83. Okuyama H, Okajima N, Sasaki S, Higashi S, Murata N (1991) Biochim Biophys Acta 1084:13
84. Henderson RJ, Millar RM, Sargent JR, Jostensen J-P (1993) Lipids 28:389
85. de Mendoza D, Cronan JE Jr (1983) Trends Biochem Sci 8:49
86. Magnuson K, Jackowski S, Rock CO, Cronan JE Jr (1993) Microbiol Rev 57:522
87. Wada H, Gombos Z, Murata N (1990) Nature 347:200
88. Wada H, Gombos Z, Murata N (1994) Proc Natl Acad Sci USA 91:4273
89. Sharp RJ, Munster KN (1986) Biotechnological implications for microorganisms from extreme environments. In: Herbert RA, Codd GA (eds) Microbes in extreme environments. Academic Press, London, p 215
90. Schlegel, HG (1993) General microbiology. Cambridge University Press, Cambridge
91. Reichardt W (1988) Microb Ecol 15:311
92. Klecka GM, Carpenter CL, Landenberger BD (1993) Ecotox Environ Safety 25:280
93. Meher KK, Murthy MVS, Goakota KG (1994) Bioresour Technol 50:103
94. Saffley LM Jr, Westerman (1992) Bioresour Technol 41:167
95. Singh L, Sai Ram M, Alam SI, Maurya MS (1995) Bull Environ Contam Toxicol 54:472
96. Halmø G, Eimhjelen K (1981) Water Res 15:989
97. Li J, Lee T-C (1995) Trends Food Sci Technol 6:259
98. Wolber PK (1993) Adv Microb Physiol 34:203
99. Wolber P, Warren G (1989) Trends Biochem Sci 14:179
100. Kajava A, Lindow SE (1993) J Mol Biol 232:709
101. Gurian-Sherman D, Lindow SE (1993) FASEB J 7:1338
102. Allmeida MJ, Pais C (1996) Appl Environ Microbiol 62:4401
103. Sode K, Nakasono S, Tanaka M, Matsunuga T (1993) Biotechnol Bioeng 42:251
104. Hikuma M, Matsuoka H, Tanaka M, Tonooka Y (1993) Anal Lett 26:209
105. Krajewska E, Szer W (1967) Eur J Biochem 2:250
106. Jones PG, Inouye M (1994) Molec Microbiol 11:811
107. Jones PG, Krah R, Tafuri SR, Wolffe AP (1992) J Bacteriol 174:5798
108. Schindelin H, Marahiel MA, Heinemann U (1993) Nature 364:164
109. Lee SJ, Xie A, Jiang W, Etchegaray J-P, Jones PG, Inouye M (1994) Molec Microbiol 11:833
110. Willemsky G, Bang H, Fischer G, Marahiel MA (1992) J Bacteriol 174:6326
111. Graumann P, Marahill MA (1994) FEBS Lett 74:157
112. Ray MK, Sitaramamma T, Ghandi S, Shivaji S (1994) FEMDS Microbiol Lett 116:55
113. Jones PG, Cashel M, Glaser G, Neidhardt FC (1992) J Bacteriol 174:3903
114. Dammel CS, Noller HR (1995) Genes Dev 9:626
115. Jones PG, Inouye M (1996) Molec Microbiol 21:1207
116. Qoronfleh MW, Debouck C, Keller J (1992) J Bacteriol 174:7902
117. Whyte LG, Innis WE (1992) Can J Microbiol 38:1281
118. Roberts ME, Inniss WE (1992) Curr Microbiol 25:275
119. Forster J (1887) Zentr Bakteriol Parasitenk Infekt Hyg 2:337
120. Hooker JD (1847) The botany of the Antarctic voyage of H.M. Discovery ships Erebus and Terror in the years 1839–1843, vol. 1, flora of Antarctica. Cramer, Weinheim

Received June 1997

Prokaryotes Living Under Elevated Hydrostatic Pressure

Daniel Prieur[1,2] and Viggo Thor Marteinsson[1,*]

[1] CNRS, UPR 9042, Station Biologique, BP 74, 29682 Roscoff. *E-mail: prieur@sb-roscoff.fr*
[2] Université de Bretagne occidentale, Brest, France

Prokaryotic life is controlled by many parameters that play a role in every colonized habitat. However, only a few biotopes are exposed to elevated pressures. That is the case for most of the oceans whose average depth is 3800 m, which corresponds to a hydrostatic pressure of 38 MPa. However, recently discovered subsurface biotopes are also exposed to high pressures. These biotopes are colonized by procaryotic organisms that display different responses to high pressures, mostly depending on their origins. Most of the studies dealing with pressure on microorganisms living in high pressure habitats concern psychrophiles (usual inhabitants of the deep-sea) and some hyperthermophiles isolated from deep-sea hydrothermal vents. The organisms so far isolated and described, their physiology and adaptations to high pressure conditions will be presented.

1
Introduction

Prokaryotic life is driven by availability of energy and carbon sources and electron acceptors, but also by numerous physio-chemical parameters. Most of

---

* Present address: Department of Biotechnology. Technological Institute of Iceland. KELD-NAHOLT, IS-112 REYKJAVIK, ICELAND.

Advances in Biochemical Engineering /
Biotechnology, Vol. 61
Managing Editor: Th. Scheper
© Springer-Verlag Berlin Heidelberg 1998

these parameters (temperature, pH, salt concentrations, redox potential, etc.) may influence prokaryote physiology in all ecosystems of the biosphere. However, some others are specific to certain habitats, and this is the case for hydrostatic pressure.

Hydrostatic pressure is a function of the weight of liquid above a given surface and is expressed in kg cm^{-2}, atmosphere, bar or MPa. Roughly, 1 kg cm^{-2} = 1 atm = 1 bar = 0.103 Mpa, and pressure increases by 0.103 MPa every 10 m of water.

The surface of the planet Earth is 70% covered by seas and oceans, with a maximum depth of 10,790 m (Marianas Trench), corresponding to approximately 110 MPa. The average depth of the oceans is 3800 m, and, accepting that deep-sea corresponds to depths greater than 1000 m, it has been calculated that this water volume corresponded to 62% of total biosphere [1]. Therefore, the deep-sea represents the largest biotope of the planet, and the most influenced by hydrostatic pressure. However, other habitats recently discovered in subsurface environments such as deep ground water [2], deep sediments [3], oil fields [4, 5], are also influenced by high pressures, combining effects of both liquid and solid materials.

The deep-sea is influenced by three major parameters which are low nutrient concentration, high hydrostatic pressure and low temperature (at the average depth of 3800 m, the mean temperature is 2 °C [6]. For these reasons, studies on pressure effects on deep-sea bacteria mostly concern psychrophiles or mesophiles. However, with the discovery of deep-sea hydrothermal vents where hot fluids remain liquid at temperatures above 350 °C because of the pressure, effects of pressure on thermophiles and hyperthermophiles were also studied [7]. Finally, in order to understand the mechanisms of bacterial adaptation to high pressures, some terrestrial organisms have also been exposed to these unusual conditions (for them) and their physiology, biochemistry and molecular biology studied. Several detailed reviews have been published on microbial life at high pressures [6, 8, 9], and the mechanism of bacterial adaptation has been well adressed recently [10]. This paper will focus on prokaryotes usually living permanently or temporarily in high pressure environments.

2
Deep-Sea Psychrophiles

2.1
Evidences for Barophilic Procaryotes

The occurrence of bacteria living in oceanic waters and sediments at depths of 5000 m was first demonstrated by Certes [11] working on samples collected during the oceanographic expedition of the "Travailleur" and the "Talisman" (1882–1883). However, after the pionneer work of Certes [11] and Fisher [12], deep-sea microbiology only developed in the second half of the century on the impulse of Zobell [13] and Morita [14]. Since the deep-sea is an oligotrophic environment, the possibility that organic matter could be decomposed under deep-sea conditions (low temperature and high pressure) was first addressed.

Heterotrophic activity of marine bacteria isolated from surface waters appeared to be "considerably retarded" under these conditions, and similar results were obtained from experiments performed with deep-sea organisms [15]. Several reviews intensively described these studies that used "*in situ*" experiments, ship-board analysis on decompressed and undecompressed samples [16, 17]. It was demonstrated that respiration of deep-sea bacteria was less affected by high pressure and low temperature than substrate incorporation, and concluded that probably no free-living bacteria adapted to deep-sea conditions existed [18–20]. But it was also pointed out that the deep-sea microbial communities could be dominated by surface-borne bacteria accompanying particles sinking from the surface [17]. Effectively, it was demonstrated that bacteria associated with surface particles lost their activity when entering deep and cold waters. Then several authors suggested that barophilic bacteria could exist within digestive tracts of deep-sea invertebrates. These proved successful first with the study of bacteria isolated from deep-sea amphipods, and were then confirmed by observations of barophilic responses to organic enrichments of mixed microbial communities from the digestive tract of benthic holothurians and fishes [21–24]. Many marine invertebrates and vertebrates migrate vertically for short periods (marine mammals), daily (plankton, fishes) or during their life cycle (crustaceans) and so their associated microflora undergo pressure variations (8). These animals also represent a favourable niche for at least barotolerant microbial communities, and sediment trap experiments showed that fecal pellets contained more barophilic bacteria when collected at increasing depths (25).

2.2
Description of Barophilic Strains

The most significant effort in searching for barophilic bacteria has been carried out by Yayanos and his group since 1978. The term "barophile" corresponds to a microorganism that requires high pressure for growth, or grows better at pressures higher than atmospheric pressure [26]. Recently, it was suggested that piezophile was more appropriated [27], but barophile still remains commonly used. The first barophilic bacterium was isolated from dead and decomposing amphipods previously captured in a pressure-retaining trap at a depth of 5600 m [28, 29]. This organism designated CNPT3 was isolated at 2 °C and 570 atm, but grew optimally between 2 and 4 °C at 500 atm, with a doubling time between 4 and 13 h. The doubling time increased by up to 3–4 days when this *Spirillum* sp. was grown at the same temperature, but at atmospheric pressure. A major step for the study of barophily was achieved with the isolation of strain MT41 [30], again from a decomposing amphipod, *Hirondella gigas*, but captured at a depth of 10,476 m (almost the maximum depth recorded). This organism did not grow under pressure below 380 atm, and grew optimally at 2 °C and 690 atm with a doubling time of 25 h. Furthermore this true barophilic behaviour was confirmed when the strain appeared to loose its ability to form colonies under in situ conditions after being exposed for several hours at atmospheric pressure. Yayanos and his co-workers isolated a set of strains from samples collected from 1957 to 10,476 m, and studied their responses to hydrostatic pressure. They came to the conclusions [31] that barophily is a general feature of bacteria from cold deep-seas, but that optimal pressure for growth at 2 °C for a given strain is always below the corresponding pressure at the depth of sample collection. Thus, although not corresponding exactly to the capture depth, the optimal pressure for a strain is an indicator of its origin. However, the pressure tolerance range varies according to the strain capture depth and is maximum for those coming from 5600 – 5900 m. Moreover, it seems that a threshold exist at 2000 m [32], with the upper layers dominated by barotolerant bacteria, and the deeper layer inhabited by more and more barophilic organisms [33]. But it must be pointed out that all these heterotrophic barophilic strains have been obtained from nutrient-rich environments (invertebrate digestive tracts or fecal pellets) while it is well established that, with a few exceptions represented by sinking corpses of big marine animals, and hydrothermal vent communities, the deep-sea is characterized by low nutrient concentrations. Also, total mineralization of organic matter requires participation of oligohetero-trophs and chemoautotrophs. Only description of barophilic activity for such organisms would indicate that barophily is an ubiquitous essential feature of deep-sea bacteria [34]. But, since autotrophs frequently grow slowly compared to heterotrophs, and barophilic psychrophilic heterotrophs have a rather long doubling time (several hours), searching for baro-psychro-autotrophs is probably very difficult.

Almost contemporary with Yayanos's studies or later on, several deep-sea microbiologists reported isolation of barotolerant and barophilic strains from various deep samples (mostly invertebrate and fishes), so confirming the occurrence of barophilic micro-organisms in the deep ocean [24, 35, 36].

All the above data concerned psychrophiles, and it is well known that these organisms are very sensitive, even to moderate temperatures. This may explain that true barophiles are rare [32], at least in the scientific reports. Also, temperature and pressure may be difficult to address separately. From experiments recently carried out with barophilic and barotolerant strains, it was reported that barophily varied for these two kinds of strains at low temperature, but was almost similar when cultured near their upper temperature limit for growth (36).

2.3
High Pressure Responses of Deep-Sea Mesophiles

Since the deep-sea is a permanently cold environment, no mesophiles were expected to be living permanently in these habitats. However, within hydro-thermal vent areas, mesophilic niches do exist. In these tectonically active regions, sea water enters cracks and penetrates some hundred meters below the sea floor [37]. Seawater thus becomes heated when approaching the magma chamber and vents out after a very complex underground circulation. If some mixing occurs between the hot fluid and cold seawater, just before venting out, the fluid is warm (10 – 30 °C). If no mixing occurs, the fluid remains very hot, and contact of the hot reduced mineral-enriched acidic fluid with cold oxygenated sea water produces precipitation of minerals that form the so-called black-smokers. But the temperature gradients are very steep, and again warm zones exist close to the smokers. Invertebrate communities thrive in these tepid environments, fed by efficient chemoautotrophic endosymbionts [6]. Besides these symbiotic prokaryotes, various free-living or surface bound bacteria exist. Several papers have been published on these microorganisms, but very few addressed the question of pressure. A mesophilic sulphur-oxidizing strain isolated from samples collected at a depth of 2000 m in the North-Fiji Basin, was described as a new species of *Thiobacillus*, *T. Hydrothermalis* [38]. This strain appeared to be barosensitive, but it must be pointed out that the culture conditions under high pressure currently used in the laboratory do not allow for a headspace for gas above the liquid medium, and that is not favourable for cells requiring oxygen for oxidation of sulphides.

Manganese is abundant in the hot fluids, and is often used as a plume tracer. Manganese-oxidizing bacteria have been reported from deep-sea vents and their influence on the scavenging and partitioning of manganese have been reported [39]. *In situ* manganese-binding experiments using these micro-organisms indicated that the process is enhanced by hydrostatic pressure, and the hypothesis of "a binding mechanism involving predominantly organic polymers produced by barophilic bacteria" has been postulated.

With the exception of barophiles reported above, many deep-sea bacteria have been isolated under atmospheric pressure conditions, which indicates that they are (at least) barotolerant. These culture conditions may have favoured non-indigenous microorganisms in the culture enrichments. For these reasons, it was attempted to enrich deep-sea bacteria under hydrostatic pressure. Comparisons of strains isolated from the same sample (a deep vent bivalve), on the same culture medium but at atmospheric and hydrostatic pressures, revealed that most of the strains obtained under the pressure conditions grew faster and exhibited more enzymatic activities when grown under high hydrostatic pressure [40]. Such mesophilic microorganisms have generation times comparable to their terrestrial or coastal counterparts and could represent additional interesting models for studies on high pressure adaptations.

2.4
Adaptations to High Pressures

The availability of barophilic bacteria in pure culture allowed experiments designed for a better understanding of the physiology of bacterial responses to pressure to be carried out. The strain CNPT3 was studied for its membrane fatty acid composition as a function of pressure [41]. Under high pressure conditions a decrease of $C_{14:1}$, $C_{16:0}$ and $C_{14:0}$, but an increase of $C_{16:1}$ and $C_{18:1}$ were observed. The ratio of total unsaturated fatty acids to total saturated fatty acids increased from 1.9 to 3 for pressures increasing from 1 to 680 atm. A similar increase of unsaturated fatty acids was reported for *Vibrio marinus*, when incubation temperatures decreased from 25 to 15 °C [42]. Eleven marine strains isolated from depths ranging from 1200 to 10,476 m were similarly studied [43]. Three of these isolates, grown at 2 °C, contained more polyunsaturated fatty acids (especially long chains up to C_{22}) with increasing pressures up to the optimum. It has been proposed that these modifications in lipid composition for cells exposed to high pressures, or low temperature, were correlated with membrane fluidity. The barophilic strains CNPT3 and PE36, and *V. marinus* were chosen for the study of PTS (sugar phosphotransferase system) that is involved in sugar transport [44]. A positive correlation for sugar phosphorylation was found between barophily and enzyme activity as a function of pressure. The effect of hydrostatic pressure on extracellular enzymatic activities has been studied by several authors. In the case of chitinase (45), it was found that chitinase synthesis was inhibited by elevated pressures (400 atm), and that chitinase activity measured on cell extracts from different strains was rather barotolerant. The modification of the commercially available API ZYME assay kit allowed for experiments with whole cells incubated in stainless-steel vessels under deep-sea conditions [46]. For nine out of ten barophiles studied by this method, at least one enzyme reaction was affected by high pressure. Pressure did not affect enzymes with a strong activity, but clearly enhanced activities of esterase-lipase, leucine aminopeptidase, β-galactosidase or *N*-acetyl-β-glucosaminidase.

2.5
Taxonomy and Phylogeny of Psychrobarophiles

Some of the barophilic organisms so far isolated have been studied by molecular taxonomy. All belong to the gamma sub-group of proteobacteria as revealed by 5S rRNA and 16S rRNA sequence analysis. Two barophiles isolated from deep-sea invertebrates collected from the Puerto Rico Trench have been assigned to the genus *Shewanella* by 5S rRNA sequence analysis, and the type strain was designed *Shewanella benthica* [47]. More recently, the 5S rRNA sequence of an obligate barophile isolated from the same area at 7410 m depth was determined, and it appeared to be similar to *Vibrio psychoerythrus*, both strains appearing distinct from the other vibrios. The new genus *Colwellia* has been created for these strains. *V. psychoerythrus* was renamed *C. psychoerythrus*, and the barophile *C. hadaliensis* [48, 49]. More recently, 16S rRNA sequences of eleven barophiles were determined, and analysis of these sequences revealed five lineages corresponding to *Colwellia*, *Shewanella*, *Photobacterium*, *Vibrio marinus* and an undescribed genus [36]. No clear correlation was found with depth, isolation or nature of the sample. While other extremophiles (extreme halophiles, extreme thermophiles) represent distinct phylogenetic lineages, barophiles so far studied appeared to be distributed in lineages which include baro-sensitive members, indicating that "high-pressure adaptation has evolved separately during relatively recent bacterial speciation".

2.6
Molecular Biology of Barophiles

Exposure of mesophilic organisms to high or low temperatures induces production of heat shock or cold shock proteins. Search for similar response to high pressure exposure was carried out for the facultative barophile *Photobacterium* SS9. This bacterium was isolated from a depth of 2500 m in the Sulu Sea. It grows optimally at 280 atm, but also at 1 atm, and at temperatures up to 23 °C.

When grown at different pressures, total protein patterns were rather similar, but under optimal pressure, one abundant protein appeared to be repressed (OmpL), and several other induced (OmpI being induced for pressures beyond SS9 optimal pressure, 400 atm). The most abundant pressure-induced protein was found to be a 37 Kda outer membrane protein called OmpH [50]. Information about OmpH function and regulation was obtained by cloning the *ompH* gene and constructing several *ompH* mutants [51, 52]. Increasing pressures up to SS9's optimum pressure increased rates of transcription of *ompH* and subsequently its transcript levels. It was found that OmpH probably "functioned as a relatively non-specific porine protein, that facilitated uptake of substrates larger than 400 daltons". Other environmental factors such as cell density or carbon starvation [53] were found to influence expression of *ompH* gene, the same promoter being "activated by high cell density at 1 atm as well as during low-cell density growth at 272 atm". Analysis of transposon mutants isolated for this study indicated that insertions occurred in the third gene of a four-gene operon. This operon showed similarities to the *rpoE* and *algU/algt* operons of *E. coli* and *P. aeruginosa*, suspected to be involved in sensing and regulating gene expression in response to modifications in the extracytoplasmic environment of bacterial cells. Among the mutants impaired in *ompH* expression isolated, one called EC1002 was also altered in induction of OmpI and repression of OmpL under high pressure (54). This mutant was found to produce less C20:5 (EPA: eicosapentaenoic acid) than the the wild strain SS9. That could indicate that Omp proteins and EPA are controlled by the same regularory gene, or that EPA or one of its precursor may regulate Omp synthesis.

Another deep-sea barophilic bacterium isolated from mud samples (strain DB 6705) was investigated at the molecular level, and a pressure-regulated promoter was cloned and sequenced [55, 56]. For this purpose, DB 6705 DNA was cloned in a plasmid vector immediately upstream of a promoter-less CAT gene. Then *E. coli* transformants expressing chloramphenicol resistance at 30 Mpa were selected. It was then shown that this pressure regulated promoter directed the same transcription start sites in *E. coli* and the barophile DB 6705. More recently, the analysis of the downstream region of this promoter revealed two open reading frames organized as an operon. However, functions of proteins encoded by these ORFs are still unknown. Two directly repeated sequences within this promoter region were found to be similar to to sequences from the promoter region for *ompH*.

3
Deep-Sea Hyperthermophiles

Although many questions about adaptation of psychrophiles and mesophiles to hydrostatic pressure remain unanswered, much more is known about pressure adaptations for psychrophiles than for their thermophilic counterparts. It must be remembered that although the first moderate thermophile, *Bacillus stearothermophilus* was described in 1920 [57], the first thermophile and hyperthermophile, *Thermus aquaticus* and *Sulfolobus acidocaldarius* were only published about in 1969 and 1972 respectively [58, 59]. Also, the deep-sea hydrothermal vents and the spectacular so-called "black smokers", venting hot fluids with temperature up to 350 °C and above, were discovered less than 20 years ago [60].

3.1
Thermophilic Life at Deep-Sea Hydrothermal Vents

These hot fluids, that remain liquid because of the high hydrostatic pressure, contain high amounts of methane, hydrogen and hydrogen sulfide. It has been suggested that part of these gases could have a biological origin [61], but this required demonstration that bacterial communities able to grow under black-smoker conditions (250–300 °C, 250 atm) exist. Reports of mixed cultures growing from 150–250 °C at 265 atm, with doubling times of 8 h, 1.5 h and 40 min at 150, 200 and 250°C respectively were published [62]. A major controversy followed these very surprising results that, despite several attempts, have never been confirmed, probably stemming from artefacts occurring during the criticised experiments [63, 64]. However, in the same year (1983), a new hyperthermophilic archaeon, *Pyrodictium*, growing optimally at 105°C was reported [65]. For the first time an organism was found which was able to grow beyond the 100 °C limit (water boiling point at 1 atm). Several kinds of experiments (cell counts, ATP measurements, radio-labelled compounds incorporation assays) were carried out to try to demonstrate occurrence of microorganisms in hot vent fluids (66). The conclusions were that "bacteria observed in hydrothermal fluids are not derived from the hot hydrothermal fluids but must originate in peripheral habitats". Analysis of black-smoker fluids for their DNA contents [67] indicated that "most of the superheated (174–357 °C) smoker fluid samples contained particulate DNA in concentration too high (0.86–1.32 ng ml^{-1} in samples containing 15–16 % maximum of sewater) to be attributable to entrained sea water. Correlations between DNA contents and total cell counts favoured the association of DNA with intact cells. However, these data could not fully demonstrate that microorganisms could live under black-smoker conditions. The black-smokers, also called hydrothermal chimneys, are very fragile mineral structures, and are frequently damaged by sampling devices. Because the velocity of hot fluids is in a range of 0.7–2.4 m s^{-1} [68], smoker debris that may contain living cells (from the inner or the outer parts of the chimney) are easily entrained by the plume.

3.2
Taxonomy of Deep-Sea Hyperthermophiles

However, hyperthermophilic microorganisms have been regularly isolated from deep-sea hydrothermal vents [69], particularly from smoker debris, but also in one case from worm tissues (in fact this worm lives on the outer wall of smokers), and from hot fluid samples (probably for the reason explained above). With the exeption of aerobic thermophiles (*Thermus* and *Bacillus*) recently reported [70, 71], and a few unpublished reports of Thermotogales, all the deep-sea thermophiles and hyperthermophiles so far described belong to the Archaea domain [69]. Members of *Desulfurococcales, Pyrodictiales, Archaeoglobales, Methanococcales, Methanopyrales,* but mostly *Thermococcales* have been described [72]. Rather surprinsingly, no novel genus specific to deep-sea vents has been found and the deep-sea strains represent novel species of already known genera: *Pyrodictium abyssi* [73], *Archaeoglobus profundus* [74], *Pyrococcus abyssi* [75], *Thermococcus profundus* [76], etc. Sometimes representatives of the same species have been found in both shallow and deep vents: *Staphylothermus marinus* (77), *Methanopyrus kandleri* (78). Although some of these species probably have an ubiquitous distribution (for instance *Thermococcus litoralis* has been also found [4, 5] in off-shore and terrestrial deep oil fields), one of the reasons for the apparent lack of taxonomic novelty for deep-sea thermophiles is that they have been isolated using conventional microbiological methods, previously designed for shallow waters. All deep-sea thermophiles obtained up to now are at least barotolerant, since they come from deep areas, but have been cultivated in the laboratory under atmospheric pressure (or a slight hyperbaric pressure to avoid boiling of the medium and oxygen intrusion into the cultures). However, several authors investigated the effect of hydrostatic pressure on the growth rate and temperature range for growth of certain deep-sea strains.

3.3
Responses to Hydrostatic Pressure

Different behaviours were reported from these experiments, carried out accord-ing to different protocols [7]. Pressure was shown to increase from 1 to 4 °C, the maximum temperature for growth of strains GB-D, GB4 [79], *P. abyssi* GE5 [75], ES1, ES4 [80]. Optimal temperature for growth was also shifted up from 2 to 4 °C for strains ES1, ES4 [80] and GE5 [75]. Finally, growth rate was increased by elevated pressure for strains AL2 [81], GE5 [75], ES1, ES4 [80], GB-D, MAR7-C, and SY [79]. All these strains are hyperthermophilic strictly anaerobic Archaea (*Desulfurococcales* and *Thermococcales*) that ferment organic compounds (mostly peptides) and respire, obligately or not elemental sulphur [72]. Effect of pressure on hyperthermophilic methanogens was also studied, with the difficulty that a CO_2/H_2 headspace above the liquid culture does not allow easily pressurization of the culture. For the strains CS 1 and FS [79], a slight pressure of 3 atm increased the growth rate, probably because of the increased availabili-ty of the CO_2/H_2 mixture due to increased dissolution. But application of a 200 atm pressure clearly decreased the growth rate of these methanogens. Effect of pressure on the deep-sea methanogen *Methanococcus jannaschii* was studied using an hyperbaric bioreactor [82]. With helium as the pressurizing gas, it was shown that pressure up to 750 atm accelerated growth and methanogenesis at 86 and 90 °C. However, above 90 °C, methanogenesis and growth were uncou-pled, and pressure increased only the upper temperature limit for methano-genesis. However, with argon or H_2/CO_2 as pressurizing gas, methanogenesis did not occurr above 86 °C. The effect of hydrostatic pressure was also studied on a thermophilic methanogen originated from shallow waters, *Methanococcus thermolithotrophicus* [83]. It was observed that growth rate was enhanced by pressure, but the temperature range was not extended.

In a review of the experiments reported on deep-sea hyperthermophiles and their responses to hydrostatic pressures, Deming and Baross [7] gave definitions for the different behaviours observed. For barosensitive strains, test pressure reduced growth rate relative to lower pressure. For barotolerant strains, growth rate was unaffected by test pressure. In the case of barophilic strains, growth rate was stimulated by elevated pressure. A fourth category, obligately barophilic, was defined for the strains whose growth was enabled by pressure, i. e. no growth was observed at lower pressure, all other conditions being equal. In that case, maximum temperature for growth was shifted up under elevated pressure, and growth for these supra-optimal temperatures was only possible under elevated pressure. However, because in that case pressure cannot be separated from temperature, the term barodependent is probably more appropriate.

In a few cases, cells of hyperthermophiles grown under pressure were microscopically examined. Elevated pressure applied to *Methanococcus thermolithotrophicus* [84] resulted in "anomalously large, elongated cells which were obviously perturbed with respect to cell division". On the contrary, cells of the deep-sea hyperthermophiles AL1 and AL2 [81] "produced under each pressure tested (from 100 to 500 atm) were notably uniform in size (1–2 μm) whereas cells in low-pressure control varied from 1 to 5 μm in diameter, with larger and more-irregular cells frequently predominant". However, no change in the protein profiles of these cells as determined by SDS-PAGE was observed according to pressure variations, as it was also noted recently for *P. abyssi* [85]. A different observation was reported for *Methanococcus thermolithotrophicus*. Incubation under 500 atm resulted in alterations of the protein pattern [84]. A series of proteins with a molecular mass in the range 38–70 kDa occurred in pressure grown cells. However, "the question of whether the observed alterations are caused by the perturbation of the balance protein synthesis and turnover or by the pressure-induced synthesis of compounds analogous to heat shock proteins remained unanswered".

After the experiments reported by Baross and Deming [62], and followed by a major controversy [63, 64], no successful attempt of enrichment of hyperthermophile under pressure has been reported. Recently, such experiments were carried out with samples collected at the mid-Atlantic ridge vent sites, which are the deepest (3500 m) explored at the moment. From these enrichment cultures, a novel species of *Thermococcus* was isolated [86]. Although common from a taxonomic and phylogenetic point of view, this organism showed a higher growth rate under pressure, but no shift of its optimal temperature for growth (85 °C). Examination of its protein pattern revealed that one proteic band was less expressed with pressure, when a second was apparently pressure induced. Microsequencing of these proteins did not allow to identify the pressure-induced protein, but the pressure-repressed protein appeared to be close to a heat shock protein previously identified for the hyperthermophile *Desulfurococcus* strain SY [87]. These results would suggest that absence of pressure would correspond to a stress situation for this deep-sea organism.

4
Conclusions

With the recent microbiological studies of subsurface biotopes, novel habitats whose inhabitants are exposed to high pressures and sometimes high temperature were discovered. It is particularly the case for deep continental oil reservoirs where *in situ* temperature was 70 °C and pressure between 50 and 160 atm. Growth of isolates (thermophilic Eubacteria and Archaea) obtained from one of these habitats was shown to be unaffected by *in situ* pressures. In their review on deep-sea thermophiles [7], Deming and Baross observed that, for most of the deep-sea strains experimentally exposed to elevated hydrostatic pressures, the optimum pressure for growth was often above the pressure at the site of collection of the original sample from which the cells were enriched and isolated. It must be remembered that for the psychrophiles, optimal pressure recorded was below the pressure existing at the depth of capture [34]. For instance, although obtained from a sample collected at 2000 m (200 atm), growth rate of *P. abyssi* strain GE5 was higher for 400 atm [75], relatively to 200 and 1 atm. These authors [7] suggested that the deep-sea smokers could represent "windows to a subsurface biosphere". If such a biosphere existed, the inhabitants should be exposed to a pressure (both hydrostatic and lithostatic) above the pressure existing on the seafloor. Whether or not the black smokers correspond to windows to an unexplored subsurface biotope, it is clear that deep biotopes inhabited by microorganisms exist and are certainly more expended than previously expected (man-made pressurized systems should be also investigated). Moreover, terrestrial hot springs could also be an access to subsurface biosphere, which could be explored from drill holes in geothermal heated areas. Before these discoveries, it had been estimated that the deep-sea below 1000 m represented 62% of the total biosphere on planet Earth. Now, with the discovery of novel biotopes, the deep-sea represents a smaller percentage of the total biosphere. However, the pressure exposed habitats become undoubtedly the most abundant on the planet. The parameter pressure has been addressed in the past only by a few microbiologists: it should not be further ignored in the near future.

5
References

1. Jannasch HW, Taylor CD (1984) Ann Rev Microbiol 38:487
2. Szewzyk U, Szewzyk R, Stenström T A (1994) Proc Natl Acad Sci USA 91:1810
3. Parkes RJ et al. (1993) Nature 371:410
4. Stetter KO et al. (1993) Nature 365:743
5. L'Haridon S, Reysenbach A-L, Glénat P, Prieur D, Jeanthon C (1995) Nature 377:223
6. Prieur D (1992) Physiology and biotechnological potential of deep-sea bacteria. In: Herbert RA, Sharp RJ (eds) Molecular biology and biotechnology of extremophiles. Blackie, Glasgow and London, p 163
7. Deming JW, Baross JA (1993) Geochim Cosmochim Acta 57:3219
8. Bartlett DH (1991) Res Microbiol 142:923
9. Bartlett DH (1992) Sci Progress 76:479
10. Bartlett DH, Kato C, Horikoshi K (1995) Res Microbiol 146:697
11. Certes A (1884) C R Acad Sci Paris 98:690
12. Fisher B (1894) Zentrlbl Bakteriol 15:657
13 Zobell CE (1946) Marine Microbiology Chronica Botanica Waltham
14. Morita RY (1976) In: The survival of vegetative microbes. Cambridge University Press, New York, p 279
15. Jannasch HW, Eimhjellen K, Wirsen CO, Farmanfaian A (1972) Science 171:672
16. Jannasch HW, Wirsen CO (1973) Science180:641

17. Jannasch HW (1979) Bioscience 29:228
18. Schwarz JR, Walker JD, Colwell RR (1974) Dev Industrial Microb 15:239
19. Schwarz JR, Walker JD, Colwell RR (1975) Can J Microbiol 21:682
20. Wirsen CO, Jannasch HW (1976) Environ Sci Tech 10:880
21. Schwarz JR, Yayanos AA, Colwell R R (1976) Appl Environ Microbiol 31:46
22. Ohwada K, Tabor PS, Colwell RR (1980) Appl Environ Microbiol 40:746
23. Wirsen CO, Jannasch HW (1986) Mar Biol 91:277
24. Nakayama A, Yano Y, Katsuhido Y (1994) Appl Environ Microbiol 60:4210
25. Deming JW (1985) Mar ecol Prog Ser 25:305
26. Zobell C E, Johnson F H (1949) J Bact 57:179
27. Yayanos AA (1995) Ann Rev Microbiol 49:777
28. Dietz AS, Yayanos AA (1978) Appl Environ Microbiol 36:966
29. Yayanos AA, Dietz AS, Van Boxtel R (1979) Science 205:808
30. Yayanos AA, Dietz AS, Van Boxtel R (1981) Proc Natl Acad Sci USA 78:5212
31. Yayanos AA, Dietz AS, Van Boxtel R (1982) Appl Environ Microbiol 44:1356
32. Yayanos AA, Dietz A S (1982) Appl Environ Microbiol 43:1481
33. Jannasch HW, Wirsen CO (1984) Arch Microbiol 139:281
34. Yayanos AA (1986) Proc Natl Acad Sci USA 83:9542
35. Deming JW, Tabor PS, Colwell RR (1981) Microb Ecol 7:85
36. Kato C, Sato T, Horikoshi K (1995) Biodiversity and Conservation 4:1
37. Baross J A, Deming JW (1985) Biol Soc Wash Bull 6:355
38. Durand P, Reysenbach A-L, Prieur D, Pace N (1992) Arch Microbiol 159:39
39. Cowen JP, Massoth GJ, Baker E T (1986) Nature 322:38
40. Prieur D, Erauso G, Llanos J, Deming JW, Baross J (1992) In:Balny C, Hayashi R, Heremans K, Masson P (eds) High presssure and biotechnology. Colloque Inserm/John Libbey Euro-text 224:19
41. Delong EF, Yayanos AA (1985) Science 228:1101
42. Oliver JD, Colwell RR (1973) Int J Syst Bacteriol 23:442
43. Delong EF, Yayanos AA (1986) Appl Environ Microbiol 51:730
44. Delong EF, Yayanos AA (1987) Appl Environ Microbiol 53:527
45. Helmke E, Weyland H (1986) Mar Biol 91:91
46. Straube WL, O'Brien M, Davies K, Colwell RR (1990) Appl Environ Microbiol 56:812
47. Deming JW, Hada H, Colwell RR, Luehrsen KR, Fox GE (1984) J Gen Microbiol 130:1911
48. Mc Donell MT, Colwell RR (1985) Syst Appl Microbiol 6:171
49. Deming JW, Somers LK, Straube WL, Swartz DG, McDonell MT (1988) Syst Appl Microbiol 10:152
50. Bartlett DH, Wright M, Yayanos AA, Silverman M (1989) Nature 342:572
51. Bartlett DH, Chi E, Wright M (1993) Gene 131:125
52. Bartlett DH, Chi E (1994) Arch Microbiol 162:323
53. Bartlett DH, Welch TJ (1995) J Bact 177:1008
54. Chi E, Bartlett DH (1993) J Bact 175:7533
55. Kato C, Smorawinska M, Sato T, Horikoshi K (1995) J Mar Biotechnol 2:125
56. Kato C, Smorawinska M, Sato T, Horikoshi K (1996) Biosci Biotech Biochem 60:166
57. Donk PJ (1920) J Bact 5:373
58. Brock TD, Freze H (1969) J Bact 98:289
59. Brock TD, Brock KM, Belly RT, Weiss RL (1972) Arch Microbiol 84:54
60. Edmond JM, Van Damm K (1983) Sci Am 248:78
61. Baross JA, Lilley MD, Gordon LI (1982) 298:366
62. Baross JA, Deming JW (1983) Nature 303:423
63. Trent JD, Chastain RA, Yayanos A A (1984) Nature 307:737
64. White RH (1984) Nature 310:430
65. Stetter KO, Kšnig H, Stackebrandt E (1983) Syst Appl Microbiol 4:535
66. Karl DM et al. (1988) Deep Sea Research 35:777
67. Straube WL, Deming JW, Somerville CC, Colwell RR, Baross JA (1990) Appl Environ Microbiol 56:1440

68. Converse DR, Holland HD, Edmond JM (1984) Earth Planet Sci Lett 69:159
69. Prieur D, Erauso G, Jeanthon C (1995) Planet Space Sci 43:115
70. Marteinsson VT, Birrien J-L, Kristjansson JK, Prieur D (1995) FEMS Microbiology Ecology 18:163
71. Marteinsson VT, Birrien J-L, Jeanthon C, Prieur D (1996) FEMS Microbiology Ecology 21:255
72. Stetter KO (1996) FEMS Microbiology Reviews 18:149
73. Pley UV et al. (1991) Syst Appl Microbiol 14:245
74. Burggraf S, Jannasch HW, Nicolaus B, Stetter KO (1990) Syst Appl Microbiol 13:24
75. Erauso G et al. (1993) Arch Microbiol 160:338
76. Kobayashi T, Kwak YS, Akiba T, Kudo T, Horikoshi K (1994) Syst Appl Microbiol 17:232
77. Fiala G, Stetter KO, Jannasch HW, Langworthy TA, Madon J (1986) Syst Appl Microbiol 8:106
78. Kurr M et al. (1991) Arch Microbiol 156:239
79. Jannasch HW, Wirsen CO, Molyneux SJ, Langworthy T A (1992) Appl Environ Microbiol 8:3472
80. Pledger RJ, Crump BC, Baross JA (1994) FEMS Microbiology Ecology 14:233
81. Reysenbach A-L, Deming JD (1991) Appl Environ Microbiol 57:1271
82. Miller JF, Shah NN, Nelson CM, Ludlow JM, Clark DS (1988) Appl Environ Microbiol 54:3039
83. Bernhardt G, Jaenicke R, Lÿdemann H-D, König H, Stetter KO (1988) Appl Environ Microbiol 54:1258
84. Jaenicke R, Bernhardt G, Lÿdemann H-D, Stetter KO (1988) Appl Environ Microbiol 54:2375
85. Marteinsson VT, Moulin P, Birrien J-L, Gambacorta A, Vernet M, Prieur D (1988) Appl Environ Microbiol 63:1230
86. Marteinsson VT et al. (1996) Thermophiles 96, Athens (USA), p127
87. Kagawa A (1995) Biochem Biophys Res Comm 241:2

Received June 1997

Proteins from Hyperthermophiles: Stability and Enzymatic Catalysis Close to the Boiling Point of Water

Rudolf Ladenstein[1], Garabed Antranikian[2]

[1] Karolinska Institutet NOVUM, Center for Structural Biochemistry, S-14157 Huddinge, Sweden. E-mail: ladenstein@eel.csb.ki.se
[2] TU Hamburg-Harburg, Technical Microbiology, D-21073 Hamburg, Germany

It has become clear since about a decade ago, that the biosphere contains a variety of micro-organisms that can live and grow in extreme environments. Hyperthermophilic microorganisms, present among Archaea and Bacteria, proliferate at temperatures of around 80–100 °C. The majority of the genera known to date are of marine origin, however, some of them have been found in continental hot springs and solfataric fields. Metabolic processes and specific biological functions of these organisms are mediated by enzymes and proteins that function optimally under these extreme conditions. We are now only starting to understand the structural, thermodynamic and kinetic basis for function and stability under conditions of high temperature, salt and extremes of pH. Insights gained from the study of such macromolecules help to extend our understanding of protein biochemistry and -biophysics and are becoming increasingly important for the investigation of fundamental problems in structure biology such as protein stability and protein folding. Extreme conditions in the biosphere require either the adaptation of the amino acid sequence of a protein by mutations, the optimization of weak interactions within the protein and at the protein-solvent boundary, the influence of extrinsic factors such as metabolites, cofactors, compatible solutes. Furthermore folding catalysts, known as chaperones, that assist the folding of proteins may be involved or increased protein synthesis in order to compensate for destruction by extreme conditions. The comparison of structure and stability of homologous proteins from mesophiles and hyperthermophiles has revealed important determinants of thermal stability of proteins. Rather than being the consequence of one dominant type of interactions or of a general stabilization strategy, it appears that the adaptation to high temperatures reflects a number of subtle interactions, often characteristic for each protein species, that minimize the surface energy and the hydration of apolar surface groups while burying hydrophobic residues and maximizing packing of the core as well as the energy due to charge-charge interactions and hydrogen bonds.

In this article, mechanisms of intrinsic stabilization of proteins are reviewed. These mechanisms are found on different levels of structural organization. Among the extrinsic stabilization factors, emphasis is put on archae chaperonins and their still strongly debated function. It will be shown, that optimization of weak protein-protein and protein-solvent interactions plays a key role in gaining thermostability. The difficulties in correlating suitable optimization criteria with real thermodynamic stability measures are due to experimental difficulties in measuring stabilization energies in large proteins or protein oligomers and will be discussed. Thus small single domain proteins or isolated domains of larger proteins may serve as model systems for large or multidomain proteins which due to the complexity of their thermal unfolding transitions cannot be analyzed by equilibrium thermodynamics. The analysis of the energetics of the thermal unfolding of a small, hyperthermostable DNA binding protein from *Sulfolobus* has revealed that a high melting temperature is not synonymous with a larger maximum thermodynamic stability. Finally, it is is now well documented, that many thermophilic and hyperthermophilic proteins show a statsistically increased number of salt bridges and salt bridge networks. However their contribution to thermodynamic and functional stability is still obscure.

Advances in Biochemical Engineering/
Biotechnology, Vol. 61
Managing Editor: Th. Scheper
© Springer-Verlag Berlin Heidelberg 1998

List of Symbols and Abbreviations

GluDH   Glutamate dehydrogenase
GAPDH   Glycerinaldehyde phosphate dehydrogenase
AOR   Aldehyde ferredoxin oxidoreductase
Tm   *Thermotoga maritima*
Pf   *Pyrococcus furiosus*
Cs   *Clostridium symbiosum*

Ta	*Thermoplasma acidophilum*
Taq	*Thermus aquaticus*
Bs	*Bacillus stearothermophilus*
Ha	*Homerus americanus* (lobster)
Ph	Pig heart
CS	Citrate synthase
NMR	Nuclear magnetic resonance
CD	Circular Dichroism
V	Volume
T	Temperature
T_m	Melting (transition) temperature
cal	Calorie (1 cal=4.185 Joule)
K	Kelvin
Å	Ångström (1 Å = 10^{-10} m)
N	Number of atoms
n_b	Number of buried atoms
A_c	Calculated surface area
A_o	Observed surface area
d	Distance
r.m.s.d.	Root mean square distance
C^α	Alpha carbon atom
Δ	Difference
G	Free energy
H	Enthalpy
S	Entropy
C_p	Heat Capacity

All other abbreviations are explained in the text

Keywords. Protein thermostability, Non-covalent interactions, Optimization, Structure analysis, Thermal unfolding, Chaperonins, Enzyme biotechnology.

1
Hyperthermophiles – Archaea and Bacteria

Hyperthermophiles constitute a group of microorganisms, discovered only about two decades ago, with an optimum growth temperature of at least 80 °C and a maximum growth temperature of around 10 °C. The majority of the presently known genera are of marine origin, some genera, however, have been isolated from continental hot springs and solfataric fields. The most common hot biotopes are associated with tectonically active zones on earth. The terrestrial biotopes are mainly solfataric fields which consist of soils, mud holes and surface waters heated by volcanic exhaustion from magma chambers below. The surface of the so-called "solfataras" is rich in sulfate and is acidic (pH ≤ 4). Marine hydrothermal systems are situated in shallow and abyssal depths and consist of hot fumaroles, springs and deep sea vents with temperatures up to 380 °C. A variety of microorganisms have been isolated from these exotic habitats such as the black smokers at the bottom of the Pacific Ocean. Nearly all these microorganisms are classified as Archaea including highly thermophilic organisms such as *Pyrococcus, Methanopyrus, Pyrobaculum*, species which grow at a temperature above 100 °C. To date, only two bacterial genera are represented among hyperthermophiles: *Thermotoga* and *Aquifex*. It has been shown by analysis of 16S tRNA that the hyperthermophiles are the most slowly evolving organisms within the archaeal and bacterial domains [1]. Hyperthermophiles may be the closest living descendants of ancestral life forms which evolved under conditions of high temperature. Therefore, hyperthermophiles are generally found close to the root of the phylogenetic tree, which suggests that they preceded their mesophilic counterparts in the course of evolution [2]. One has to have in mind, however, that the location of branching points in the phylogenetic tree is still hotly debated.

The maximum growth temperature which has been observed for a hyperthermophile (*Pyrodictium occultum*) is around 110 °C [3]. It is, up till now, still unknown whether this temperature represents the upper limit. It should be noted, however, that some of the building blocks of proteins, the amino acids arginine, cysteine, aspartate, glutamate undergo hydrothermal decomposition which becomes significant at temperatures around 120–130 °C [4]. Proteins isolated so far from hyperthermophiles show considerable thermal stability and seem to become increasingly important for the investigation of fundamental problems in structure biology such as the stability and folding of proteins. Extremely thermostable enzymes are more frequently used in biotechnological applications that require proteins with increased stability at high temperatures (see Sect. 6). When studying hyperthermophilic proteins, two fundamental questions arise unavoidably: firstly, what energetic and kinetic mechanisms are responsible for the thermotolerance of proteins which are adapted to function at temperatures close or above the boiling point of water, where their mesophilic counterparts would usually undergo irreversible denaturation; secondly, is there any need, to expect new forms of weak interactions or protein structures and structural elements not typical of "normal" proteins? Aspects of both these issues will be discussed in detail below.

1.1
Hyperthermophilic Proteins – Amino Acid Composition and Sequences

Referring first to the second question, we know from quite a number of biochemical studies on hyperthermostable proteins and their counterparts from mesophilic organisms, that such proteins are composed of the common twenty amino acids – thus all the information needed to create high thermotolerance is encoded in the gene sequence – and that thermotolerance is an intrinsic property. This can be deduced from the few examples where recombinant proteins expressed in *E. coli* have been obtained showing the same stability and behaviour as the wildtype proteins. Their size, complexity and functional properties are comparable to the analogous proteins from conventional sources, and for those proteins with known amino acid sequence there is usually a high degree of (sequence and! structure) homology between the mesophilic and thermophilic versions. Thus it should in principle be possible to convert a protein which denatures already around 50 °C to one which is stable at 90 °C for hours by specific mutation of amino acids in the sequence of that protein. The necessary changes, however, are not readily apparent from sequence comparisons of mesophilic and thermophilic proteins. From such studies it seems to be quite clear that specific rules for "thermostability" or generalizations can hardly be derived by systematic analyses of the sequences of homologous mesophilic and thermophilic proteins [5, 6].

Nevertheless, certain amino acid changes do occur and seem to be characteristic for thermostable proteins. The effect of hydrophobic substitutions in the protein core and their influence on stability has been investigated by site-directed mutations [7] in great detail. Substitutions of isoleucine with valine residues resulted in less stable mutants: the average effect on the free stabilization energy was $\Delta\Delta G = -1.3 \pm 0.4$ kcal mol^{-1}. The decrease of the free energy of unfolding, however, was not only a result of the reduced hydrophobic contact surface but also dependent on the created cavity volume (see below). In a comparison of the sequences of Glutamate dehydrogenases from *Pyrococcus furiosus* ($T_m = 105$ °C), *Thermotoga maritima* ($T_m = 95$ °C) and *Clostridium symbiosum* ($T_m = 55$ °C) it has been found that Pf GluDH contained 35 isoleucines, whereas only 21 and 20 isoleucines were found in the sequences of Tm and Cs GluDH, respectively. Eleven of the 35 isoleucines in Pf are conserved in Tm; 10 of them resulted from valine to isoleucine mutations and 6 from leucine to isoleucine mutations [8]. The structure-based sequence comparison on the three GluDHs has furthermore pointed out that the number of glycines is reduced drastically in the two hyperthermostable enzymes: the sequence of Cs GluDH contains 48 glycine residues; the sequences from Tm and Pf contained only 39 and 34 glycines, respectively. It is known that high glycine content may contribute to a diversity of native conformations of secondary structures. The observed reduction of glycine residues suggests a decrease of the flexibility of certain structure elements in the hyperthermostable proteins. In a recent structure comparison of ferredoxins, however, it has been found, that residues in strained conformations of the mesophilic members are exchanged with glycines in the thermophilic members [9].

1.2
Proteins are Only Marginally Stable, but the Forces Holding Them Together are Highly Cooperative

Referring now to the first question raised above, several well known facts of protein stability must be taken into account: the free energy of stabilization of a globular protein – it may be mesophilic or thermophilic – is quite small. It is not larger than about 7–15 kcal/mol [5, 10] which is equivalent to the energy which results from the formation of a few hydrogen bonds, ion pairs or hydrophobic interactions. Hyperthermophilic proteins are therefore not expected to differ greatly from mesophilic proteins; they do not exhibit properties which are qualitatively different from 'normal' proteins. The energetic stabilization of a protein is due to the cumulative effect of a large number of attractive and repulsive interactions at different locations within the molecule. Their contributions to the free energy term are quite large, if taken into account separately.

Their actual superposition in a protein, however, yields only a small difference between big numbers and thus only a marginal stabilization free energy which is attributable to the equivalent of a few non-covalent interactions. The average contributions of a hydrogen bond and a surface ion pair have been estimated to be -1.6 kcal/mol and -1.0 kcal/mol, respectively [11, 57]. Whereas equilibrium studies on chymotrypsin have indicated that the free energy of formation for a buried ion pair is roughly -3 kcal/mol [12]. However, the overall stability of a protein with a stability increment per residue, which is one order of magnitude below the thermal energy kT, can only be explained by taking into account high cooperativity of the interactions [5]. Basic stabilization mechanisms, common for all proteins, are: optimization of atomic packing equivalent to the mini-mization of the relation surface/volume, optimization of the charge distri-bution, and minimization of the accessible hydrophobic surface area. From thermodynamic stability measurements it follows that no dramatic changes in stabilization energies can be expected, not even for proteins from the most extreme environments. Rather, a decrease and broadening of the stability maxi-mum, which shifts the T_m to higher values, seems to be characteristic at least for hyperthermophilic small proteins (see below).

In conclusion, it seems to be quite obvious, that molecular adaptation to ex-treme temperatures depends in the first line only on mutational changes of the local and global distribution of the amino acids in the sequence of a protein molecule. However, in the second line, stabilization by extrinsic factors, like compatible solutes, the high protein concentration in a cell and the action of chaperonins seem to be of considerable importance, too. There is a third line of importance, suggested by the observation that a charge-charge interaction, for example can also be repulsive, i.e. destabilizing. Its actual strength, depending mainly on the distance of the charges, the solvent accessibility and the elec-trostatic field produced by the overall charge distribution, can be a target for optimization in a protein [13].

2
Protein (Thermo)Stability and Energetic Optimization of Weak Interactions

The adaptation of a protein molecule to its environment requires optimization of internal non-covalent interactions and protein-solvent interactions. The con-cept of optimization can be successfully applied to the description of thermost-ability and can reflect in an indirect way the result of the adaptation process, controlled by environmental and functional constraints, which has occured during evolution of a protein molecule.

The main question, namely, which properties are responsible for the shift in denaturation temperature of thermostable proteins, has still to be answered. However, data from structural comparisons and thermodynamic measure-ments have accumulated and can shed more light on this fundamental question. Theoretical and experimental analyses of proteins have shown that the 3D structure of a folded protein and the resulting properties, including thermosta-bility, are a result of the delicate balance of different interactions: van der Waals interactions, hydrogen bonds, charge-charge interactions, protein-solvent interactions (hydrophobic effect). Comparisons of amino acid sequences [14] and of 3D structures have suggested that thermal stability is largely achieved by an additive series of small improvements (i.e. optimization) at many locations in the macromolecule without significant changes in the tertiary structure. Their overall effect is to increase the fraction of buried hydrophobic residues, to optimize charge-charge interactions by removal of repulsive interactions and possibly organizing them in networks and to improve packing by decreasing the ratio of surface area to volume [15].

The optimization of charge-charge interactions can be described by a dimensionless parameter which has been obtained by the comparison of the contribution of charge-charge interactions to the free energy of the native protein, ΔG_{ei} (native) (see Eq. 1) and of a large number of randomly distributed charge constellations (i.e. reference state), ΔG_{ei} (random), calculated by Monte Carlo simulation [13, 16].

$$\Delta G_{ei} \text{(native)} = 1/2 \sum_i \sum_j Q_i Q_j w\,(r_{ij})\;(i, j = 1, ..., N), \tag{1}$$

(where $w(r_{ij})$ is the potential function of electrostatic interactions between the charges Q_i and Q_j separated by a distance r_{ij}, and N is the number of charged groups).

The calculations were performed on a set of 141 non-homologous proteins available in the Protein Data Bank and some selected 3D structures of hyperthermostable proteins [17]. The frequencies, $f(\Delta G_{ei}$ (random)), of a random charge constellation in the intervals of ΔG_{ei} follow a normal (Gaussian) distribution (Fig. 1), $p(\Delta G_{ei})$.

A dimensionless statistical criterion was introduced to describe the spatial optimization of the electrostatic interactions, which allows the comparison of proteins that have evolved under different environmental conditions on a common scale. The optimization parameter S_{opt} is similar to an expression used in statistics to calculate probabilities from Gaussian distributions and is defined as follows

$$S_{opt} = (\Delta G_{ei} \text{(native)} - \,<\Delta G_{ei} \text{(random)}>)/\sigma \tag{2}$$

where ΔG_{ei} (native) is the electrostatic energy of the native protein structure, $<\Delta G_{ei}$ (random)$>$ is the meanvalue of the corresponding energies of N random charge distributions, and σ is the standard deviation. S_{opt} represents a measure for the "energetic distance" of the native structure from an artificial structure with negligible contribution of the electrostatic interactions to the energetic stabilization. Increasingly negative values of S_{opt} are indicative of the structures with better optimized charge-charge interactions and vice versa (see Tables 1 and 2).

Protein-solvent interactions, in particular the hydrophobic effect, are considered to be important factors responsible for protein stability and represent presumably the most important part of the driving force for protein folding [18, 19]. Privalov and Makhadatze [20] have defined the hydrophobic interactions as a sum of the van der Waals interactions between the apolar atoms and the hy-

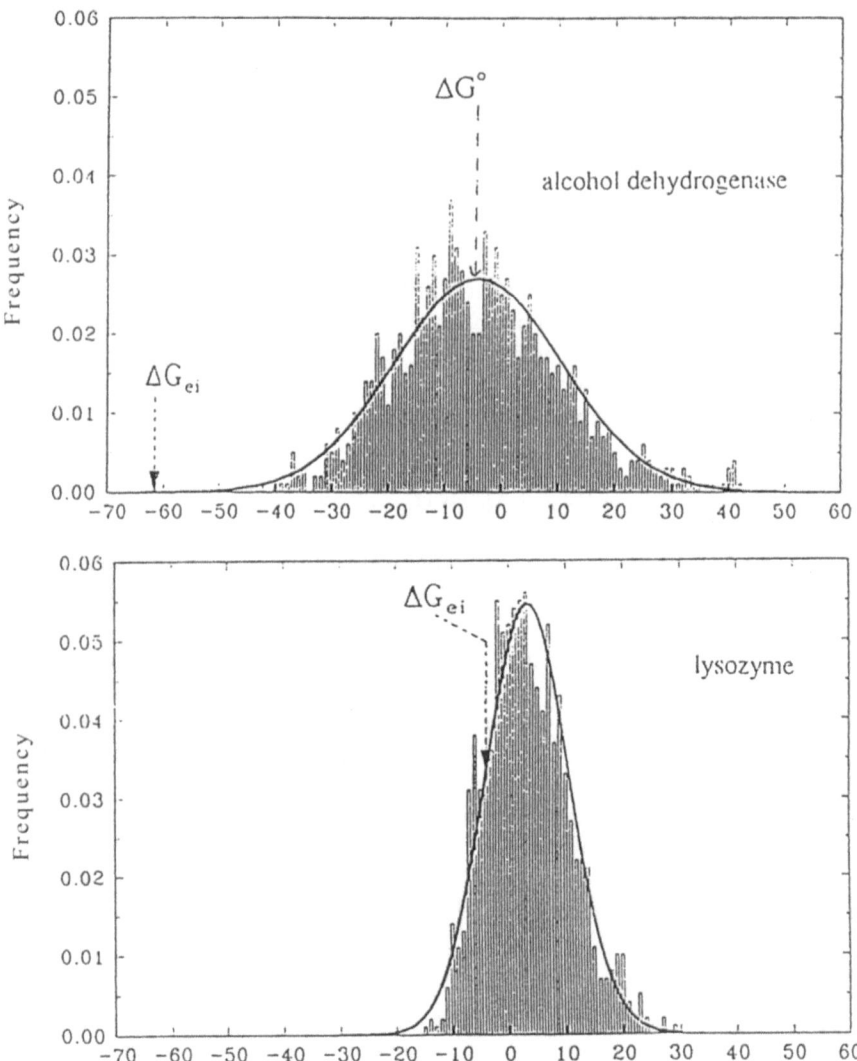

Fig. 1. Distribution of the electrostatic free energy term of interactions between randomly positioned ionized groups for alcohol dehydrogenase and hen egg lysozyme; *Vertical bars*: $f(\Delta G_{ei})^{rnd}$, computed normalized frequency of occurrence of random charge constellations as a result of Monte-Carlo simulations: *Solid line*: $p(\Delta G_{ei})$, probability density of normal (= Gauss) distribution; ΔG_{ei}, computed energy of charge-charge interactions in the native protein structure; $\Delta G°$, reference charge constellation (mean value); *Abscissa*: $(\Delta G_{ei})^{rnd}$, electrostatic free energy of random charge constellations given in kcal mol^{-1} ; (figure taken from [16])

Table 1. Optimization parameters ξ_h and S_{opt} calculated for proteins from thermophilic and hyperthermophilic organisms (*underlined*) and corresponding structures from mesophiles[a]

Protein and source	PDB code	N_{at}	ξ_h	S_{opt}	N_{slt}
Rubredoxin					
Pyrococcus furiosus	1caa	413	0.374	– 1.66	4 (4.2)
Desulfovibrio desulfuricans	6rxn	358	0.369	– 0.32	0 (1.1)
Desulfovibrio vulgaris	7rxn	389	0.380	– 0.21	1 (3.4)
Desulfovibrio gigas	1rdg	398	0.370	– 0.78	1 (3.9)
Clostridium pasteurianum	5rxn	422	0.347	– 0.50	2 (3.6)
Ferredoxin					
Bacillus thermoproteolyticus	2fxb	612	0.325	– 2.91	5 (3.1)
Clostridium acidiurici	1fdn	380	0.372	– 2.17	2 (1.0)
Desulfovibrio gigas	1fxd	430	0.323	– 2.19	1 (1.5)
Spirulina platensis	3fxc	732	0.346	– 1.46	3 (2.6)
Azotobacter vinelandii	5fd1	841	0.294	– 2.42	8 (6.6)
Thermitase					
Thermoactinomyces vulgaris	1thm	2003	0.188	– 2.88	13 (4.2)
Subtilisins:					
Carlsberg (*Bacillus licheniformis*)	1sca	1920	0.213	– 3.27	9 (3.8)
BL (*Bacillus lentus*)	1st3	1888	0.201	– 2.70	8 (3.6)
BPN (*Bacillus amyloliquefaciens*)	2st1	1938	0.219	– 3.17	10 (4.2)
Phosphoglycerate kinase					
Bacillus stearothermophilus	1php	3008	0.223	– 4.39	32 (30.4)
Saccharomyces cerevisiae	3pgk	3148	0.304	– 1.04	8 (20.8)
Superoxide dismutase (Mn, Fe)					
Thermus thermophilus	3mds	3282	0.236	– 2.02	10 (10.1)
Pseudomonas ovalis	3sdp	2910	0.287	– 1.89	6 (6.2)
Escherichia coli	1isb	3006	0.245	– 1.22	4 (6.6)
Phosphofructokinase					
Bacillus stearothermophilus	3pfk	9444	0.182	– 0.58	15 (15.2)
Escherichia coli	2pfk	9024	0.288	– 0.88	10 (15.8)
Malate dehydrogenase					
Thermus flavus	1bmd	5064	0.209	– 2.09	17 (16.7)
Porcine	4mdh	5106	0.228	– 2.57	15 (17.7)
Lactate dehydrogenase					
Bacillus stearothermophilus	1ldn	9968	0.153	– 2.50	21 (15.4)
Lactobacillus casei	1llc	9816	0.196	– 0.64	12 (14.2)
dogfish	6ldh	10180	0.194	– 2.63	19 (14.6)
Porcine	9ldb	10360	0.173	– 2.06	13 (13.0)
D-**glyceraldehyde-3-phosphate** **dehydrogenase**					
Pyrococcus furiosus	1hdg	5196		– 2.73	23 (19.5)
Thermus aquaticus	1cer	10113	0.183	– 2.55	20 (18.7)
Bacillus stearothermophilus	1gdl	10276	0.165	– 2.39	16 (19.5)
Homarus americanus (Lobster)	1gpd	10128	0.244	– 3.13	15 (13.3)
Glutamate dehydrogenase					
Pyrococcus furiosus		19752	0.176	– 4.52	42 (25)
Thermotoga maritima		18984	0.188	– 3.75	30 (24)
Clostridium symbiosum		20760	0.178	– 4.16	30 (21)

Table 1 (continued)

Protein and source	PDB code	N_{at}	ξ_h	S_{opt}	N_{slt}
Ribosomal protein					
Thermus thermophilus	1ris	817	0.292	– 2.44	9 (6.8)
Indole-3-glycerolphosphate synthase					
Sulfolobus solfataricus	1igs	2003	**0.209**	– 2.55	21 (21.1)
Thermolysin					
Bacillus thermoproteolyticus	3tln	2432	**0.208**	**	**
Elongation factor Tu					
Thermus aquaticus	1eft	3175	0.226	– 3.62	36 (28.3)
3-isopropylmalate dehydrogenase					
Thermus thermophilus	1ipd	5180	0.217	– 4.56	29 (19.5)
seryl-TRNA synthase					
Thermus thermophilus	1set	6746	0.233	– 3.78	36 (28.6)
Aldehyde ferredoxin oxidoreductase					
Pyrococcus furiosus	1aor	9386	**0.164**	– 5.36	58 (48.9)

[a] The values in bold indicate the optimization parameter calculated for thermostable proteins, if they are best optimized for a group of functionally homologous proteins. N_{slt} in the last column is the number of salt bridges in the native structure; the values given in parentheses are the number of salt bridges statistically expected for that protein structure (table modified from [23]).

dration of these atoms; the latter destabilizes the native structure due to a negative entropy term which results from the ordered water structures being created around the surface of apolar groups. Since protein stability depends on the extent of exposure of apolar groups to the surrounding solvent [21, 22] the solvent accessibility of apolar atoms has been used as an appropriate tool to analyse the contributions of hydrophobic interactions as a part of the protein-solvent interactions and their role for the stabilization of proteins [23]. It is a great advantage that solvent accessibilities of apolar groups are linearly related to the energetics of the protein [24] and can easily be calculated if the 3D structure is known. The solvation free energy term can be expressed by the equation [25]

$$\Delta G_s \text{(native)} = \sum_i \Delta f_\tau sa_i (R_{native}) \tag{3}$$

(Δf_τ is the atomic solvation parameter for atoms of type τ and has the dimension [energy/Å^2], it is positive for apolar atoms and negative for polar atoms. The atomic solvent accessibility sa_i (R) depends on the coordinates R of the individual atom i)

As a reference state the solvent accessibility of the completely unfolded protein has been used. This is justified as long as relative, dimensionless optimization criteria are considered, even in the case when we do not know too much about the structural nature of a protein in the unfolded state.

$$\Delta G_s \text{(unfolded)} = \sum_i \Delta f_\tau < sa_i (R_{unfolded}) > \tag{4}$$

Table 2. Comparison of the hydrophobic and electrostatic optimization in proteins from thermophiles

PDB code	$\delta\xi_h$	S_{opt}	Hydrophobic[a]	Electrostatic[a]
1LND	− 0.28	− 2.50	O	M
1THM	− 0.28	− 2.88	O	M
1GDI	− 0.23	− 2.39	O	M
1CAA	− 0.23	− 1.66	O	M
2FXB	− 0.17	− 2.91	O	M
3TLN[b]	− 0.17	−	O	−
1RIS	− 0.16	− 2.44	O	M
3PFK	− 0.15	− 0.58	O	L
1BMD[c]	− 0.07	− 2.09	M	M
1PHP	− 0.07	− 4.39	M	O
1EFT	− 0.05	− 3.62	M	O
1IPD	− 0.03	− 4.56	M	O
3MDS[c]	− 0.01	− 2.02	M	M
1SET	+ 0.06	− 3.78	M	O

[a] In the last two columns an evaluation of the optimization parameters is given as follows: O, high optimization: $\delta\xi_h < -0.1$, $S_{opt} < -3.0$; M, moderate optimization: $-0.1 < \delta\xi_h < 0.1$; $-3.0 < S_{opt} < -1.0$; L, low optimization: $\delta\xi_h > 0.1$; $S_{opt} > -1.0$.

[b] The electrostatic interactions for this protein are strongly influenced by specific calcium binding, therefore S_{opt} was not calculated (Spassov et al., 1994).

[c] Superoxide dismutase and malate dehydrogenase are characterized by moderate optimization of both hydrophobic and electrostatic interactions, i.e., they cannot formally be distinguished from the non-thermostable proteins. However, among the homologous proteins from mesophilic organisms, they are the best optimized structures (see Table 1; Table 2 taken from [23]).

($<sa_i (R_{unfolded})>$) is the solvent accessible area averaged over all possible random coil conformations, calculated from model peptides in extended conformation.

A dimensionless parameter, ξ_τ, has been defined as the optimization criterion for protein-solvent interactions of the atoms of type τ by

$$\xi_\tau = \sum_i sa_i (R_{native})/\sum_i < sa_i (R_{unfolded})> = SA_\tau (native)/SA_\tau (unfolded). \quad (5)$$

Different atom types can be considered by this optimization criterion: e.g. hydrophobic (apolar) atoms, characterized by $\Delta f_\tau > 0$ and ξ_h or polar atoms, characterized by $\Delta f_\tau < 0$ and ξ_p; for the hydrophobic atoms $\xi_h = 0$ corresponds to the most optimized (but not native!) structure, which is an arbitrary micelle-like structure with all hydrophobic atoms isolated from the solvent by a shell of hydrophilic atoms. According to Privalov's definition of hydrophobic interactions, the most optimized structure would correspond to a minimized hydration of the apolar atoms (due to minimization of the apolar surface fraction), which is a destabilizing factor, and maximized van der Waals interactions between these atoms in the core of the protein.

The analysis of both optimization factors revealed clearly that electrostatic as well as hydrophobic interactions are correlated with the thermal stability of proteins in a way probably obeying some common principles.

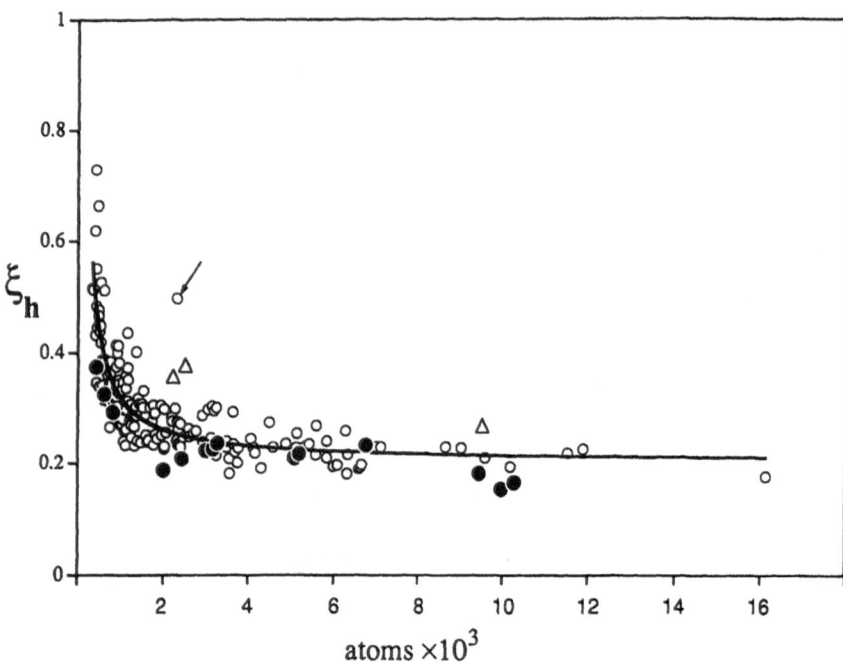

Fig. 2. Optimization parameter ξ_h versus number N_t of nonhydrogen atoms in proteins; *Open circles*: values for the representative data set; *Filled circles*: values for thermostable proteins; *Open triangles*: values for three membrane proteins 2POR, 1PRC, 3BLC; the larger ξ_h values are indicative of increased exposure of apolar groups; *Solid line*: fitted function $\xi_h(N_t) = 117.3/N_t + 0.202$, correlation coefficient r = 0,835; (figure taken from [23])

Figure 2 shows that the optimization ξ_h for the majority of thermostable proteins falls under the regression curve defined by $\xi_h(N) = 117.3/N + 0.202$ and forms the bottom limit of the distribution determined by the proteins from the representative data set. This tendency is not observed for ξ_p describing the surface fraction of the polar atoms. Thus, in terms of the optimization of protein-solvent interactions, the main contribution to enhanced thermostability comes from an increased optimization of the hydrophobic interactions. In structural terms, thermostable proteins show a tendency to minimize the solvent exposed surface fraction of apolar atoms corresponding to the minimization of the (destabilizing) hydration of these atoms. The evaluation of the sensitivity of the optimization parameters showed that a deviation from the regression curve (see above) of about 10% can be associated with changes in amino acid composition and sequence leading to a significant change in protein-solvent interactions. Proteins with $\delta\xi_h$ < 0.1 are assumed to be significantly optimized. The comparison of both $\delta\xi_h$ and S_{opt} for thermostable proteins frequently shows complementarity of the optimization of charge-charge and hydrophobic interactions. In hyperthermostable proteins there is a clear tendency for a high optimization of both parameters.

In Table 2, optimization parameters for some pairs of thermophilic and hyperthermophilic proteins are compared with functionally homologous proteins from mesophiles (psychrophiles). It is obvious that among the homologous proteins those from thermophiles are characterized by an improvement of at least one of the two optimization criteria. Interestingly, the available hyperthermophilic structures almost always show high optimization of both types of interaction. Electrostatic interactions seem to be sensibly correlated with the thermostability of small proteins as well as large enzymes molecules: the number of salt bridges is larger than the statistically expected number for the majority of thermostable proteins and in particular for hyperthermostable proteins. In particular, enzymes from hyperthermophiles such as GAPDH or GluDH show a tremendously increased number of salt bridges when compared with the mesophilic counterparts. However, the high optimization of both parameters in the case of GluDH from the mesophile *Clostridium symbiosum* is presently not understood and needs to be explained [17]. It cannot be excluded, that the high optimization of the mesophilic counterpart is due to another still unknown functional, biological or environmental constraint.

Both procedures based on optimization of electrostatic and hydrophobic interactions described above have some similarity in the generality of the approach in common, they delineate and compare global and fundamental properties of proteins which are related to "thermal stability" in the way how people frequently use this term upon comparing melting temperatures of proteins or temperature optima of enzyme activities. However, the correlation of these parameters with real thermodynamic stability measures (e.g. the free energy function ΔG) is to date unproven due to experimental difficulties of measuring free stabilization energies in large proteins (see below). The energetic basis for a correlation of optimization parameters, if any, with thermodynamic stability remains thus still unclear. One of the related problems is that proteins are multifunctional systems and nature (evolution) seems to favour optimum function rather than maximum stability. Optimizing a multifunctional system requires compromises [26].

2.1
Another Target for Optimization: The Ratio [Surface/Volume]

Aldehyde ferredoxin oxidoreductase (AOR) from *Pyrococcus furiosus* is an example of a hyperthermostable protein with a relatively small solvent-exposed surface area and a relatively large number of both ion pairs and buried atoms. The crystal structure of this homodimeric enzyme (M = 133 000) has been recently determined at 2.3 Å resolution by multiple isomorphous replacement and electron density averaging between multiple crystal forms [27]. Each of the two identical subunits contains an Fe_4S_4 cluster and a molybdopterin based tungsten cofactor that is analogous to the molybdenum cofactor found in a large class of oxotransferases which catalyze the transfer of an oxo group to or from a substrate molecule in a two electron redox reaction [28]. Many enzymes utilize first-row transition elements at their active sites, but only two second- or third row transition metals, tungsten and molybdenum, show such a biological role. Tungstoenzymes have up till now been found mainly in thermophilic anaerobes and function, in the case of AOR, in the conversion of aldehyde groups to carboxyl groups.

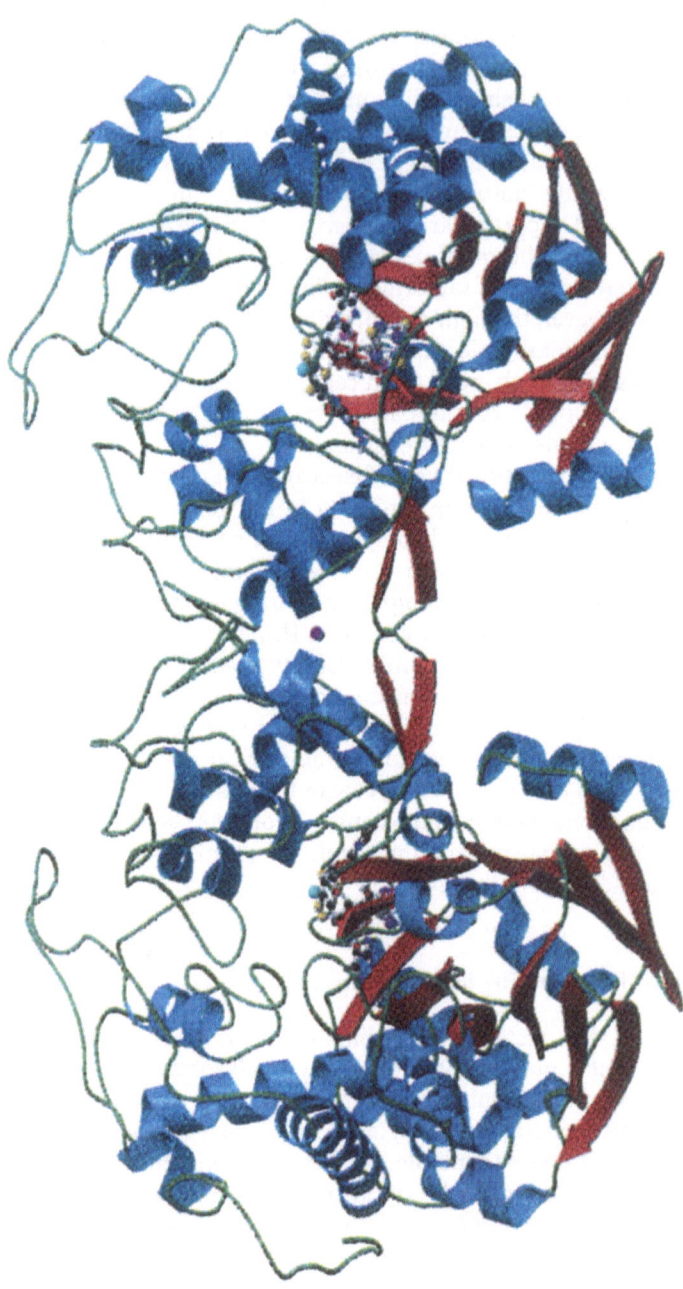

Fig. 3a. Structure of aldehyde ferredoxin reductase (AOR) from *Pyrococcus furiosus*; Ribbon diagram of the AOR dimer viewed perpendicular to the twofold axis, α helical regions: *light blue*, β sheet regions: *dark red*, loop regions: *green*; Atoms in the cofactors are colored by elements: *C black, N blue, O red, S yellow, Mg orange, Fe purple, W cyan, phosphate magenta*

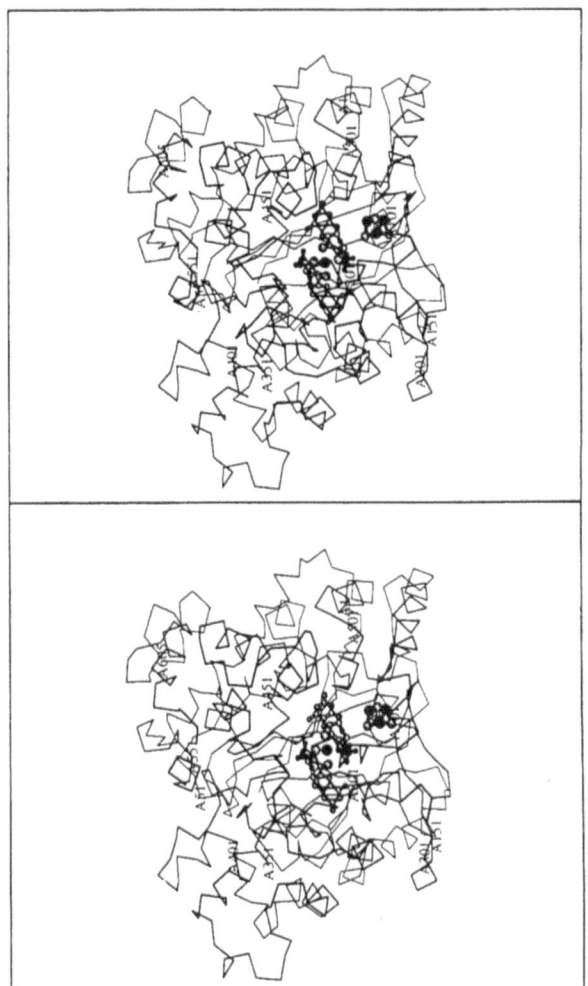

Fig. 3b. Stereo diagram of the polypetide chain of a single AOR subunit; (figure taken from [27])

AOR is the first hyperthermophilic enzyme that has been structurally characterized (Fig. 3). In the absence of a mesophilic counterpart for direct comparison the authors have based their analysis of the origins of thermal stability on
general features of protein structure established from mesophilic structures.
The primary, secondary, tertiary structures of AOR are "normal" for water-
soluble, globular proteins. The amino acid composition is close to the average
for prokaryotic proteins [27]. The secondary structure content of AOR which
has been reported (45% α helix and 14% β sheet) is within the limits seen for
globular proteins, and the packing interactions on the tertiary structure level in
between helices and strands are also "standard". AOR contains a relatively high
number of saltbridges (d < 4.0 Å) on a per residue basis, however in other proteins a similar percentage of stabilizing ionic interactions can be found. Thus it
has been suggested that ionpairs might be of importance, but cannot represent
the sole determinant of thermostability in AOR. One might ask, what is important then?

The application of the concepts of molecular surface area and volume for the
analysis of macromolecular structures has been pioneered by Richards [28].
The picture of a water-soluble, globular protein which emerges from this work
points out i) a smooth surface that minimizes the surface energy in an aqueous
environment, and ii) the efficiently packed, relatively apolar core. Calculations
of the solvent accessible surfaces have shown that the AOR dimer was expected
to have about a 17% larger accessible surface area on the basis of a survey of
water-soluble oligomeric proteins [30]. The surface of AOR appeared to have
reduced area, but comparable polarity when compared to other globular protein structures.

That the minimization of the surface area/volume ratio might indeed contribute to the thermostability of AOR has been computationally assessed by evaluation of the (dimensionless) relative surface area A_o/A_c and the number of
buried atoms, n_b, for a representative set of protein structures [27]. A_c, the expected area for a generic protein has been obtained from the fitted equation
$A_c = 15.0 \, N^{0.866}$. A_o/A_c provides an estimate for the extent of surface area minimization of a given protein.

Evaluation of the packing efficiency is more difficult and complex, but as a
simple indicator the fraction of buried atoms, n_b, with zero accessible surface
area has been used. The value of n_b may be obtained directly from surface area
calculations. The larger the value of n_b the larger the extent of atom sequestra-
tion from solvent exposure. It has been pointed out that, in terms of these parameters, low values of A_o/A_c and high values of n_b indirectly show a minimization of the ratio surface area/volume of a protein. A plot of A_o/A_c vs n_b has indicated negative correlation for both values and, in addition, that the AOR
structure simultaneously showed both the minimum value for the relative surface area ratio and the maximum fraction of buried atoms [27]. This result has
emphasized that AOR has a reduced ratio of surface area to volume relative to
the other 30 mesophilic proteins of the representative data set. However a drawback of the method is the extremely indirect evaluation of the volume, a quantity which indeed cannot be obtained in a straightforward way for protein molecules. A conclusion from the results discussed above has been that these effects,

which are due to distinct structural properties, may be related to the enhanced thermal stability of AOR.

2.2
"Buried" Volumes: Cavities and Packing

The minimization of the ratio surface area/volume increases the stability of an object by simultaneously reducing the unfavorable surface energy and increasing the attractive interior packing interactions. Although the effects of surface area variation have not been systematically explored, the influence of variations in packing interactions, in particular through the formation or removal of cavities has been more extensively documented [31]. The introduction of cavities into a protein is often associated with decreased stability. A study of the sizes and distributions of cavities in 12 proteins has indicated that cavities are quite common and sometimes large [32]. The unfavourable energy contribution of a cavity has been estimated to be on the order of $25-60$ cal mol^{-1} Å^3 [33]. A large number of mutation studies have highlighted the importance of the role of cavities for protein stability [31]. Cavity-creating mutations have consistently resulted in proteins with decreased thermodynamic stability with respect to the wild-type [33]. Alternatively, in some cases, mutations have been shown to cause an increase in the stability of the mutant [34].

Thermoplasma acidophilum (Ta) citrate synthase is a dimer of about 90,000 Da. The gene has been cloned and expressed in *E. coli* [35, 36]. The enzyme is active after incubation for 10 min at 78°C and no change in helix content has been observed in CD spectra recorded at 80°C. Pig heart (Ph) citrate synthase shows no activity after incubation for 10 min at 45°C [37]. The Ta enzyme exhibits only 20% sequence identity with dimeric Ph citrate synthase. Both enzymes nevertheless share a high degree of structural homology. The crystal structures of citrate synthase from pig and chicken heart have been determined [38]. The crystal structure of the "open" form of Ta citrate synthase has been determined at 2.5-Å resolution by Patterson search methods with the pig enzyme as the search model and subsequently been improved by crystallographic refinement [39].

The compactness of thermostable citrate synthase from *Thermoplasma acidophilum* has been compared with the mesophilic counterpart from pig heart [39]. Detection of cavities have been performed using the program VOIDOO [40], usually with a 1.4-Å probe radius and a 0.75-Å grid and a multirotational approach on order to reduce errors caused by the grid specification. The cavity calculations have been carried out on the dimers of both citrate synthases. Calculations detecting probe accessible cavities have revealed 11 cavities ($V = 96$ Å^3) in pig heart citrate synthase and only seven in the thermostable citrate synthase ($V = 29$ Å^3). The tightened atomic packing apparent in *Thermoplasma* citrate synthase has been discussed as a factor responsible for its increased thermostability. The volume difference of the cavities in both enzymes is $\Delta V = 67$ Å^3, resulting in a stabilizing free energy increment for *Thermoplasma* citrate synthase of $1.6-4.0$ kcal mol^{-1} at room temperature by using the estimation of Eriksson et al. [33].

A reduction of the probe accessible cavity volume has also been observed in the comparison of GluDHs from Cs, Tm and Pf; cavity volumes per monomer of 107.6 Å^3, 45.1 Å^3 and 21.3 Å^3, respectively, have been detected [8]. In a more extended analysis of cavities, voids and partial specific volumes and its implication for protein thermostability a tendency of cavity volume reduction has not been confirmed for thermostable proteins. However, a clear trend has been observed when enzymes belonging to the same structural and functional families were compared [41].

Whether or not the cavity volume in these enzyme changes at the growth temperature of the host remains unclear and indicates an important problem, which is often neglected in our way to look at hyperthermostable structures: the 3D structural coordinates which are used to delineate molecular properties are generally obtained from X-ray or NMR structure determinations at temperatures of 20°C or at even much lower temperatures, when the crystals are cooled during diffraction measurements. Whether or not molecular properties corresponding to hyperthermostability may change significantly at the temperature optimum of the host relative to room temperature, remains to be determined. X-ray diffraction measurements of hyperthermostable enzymes at increased temperatures, although experimentally difficult to realize, are therefore highly desired and necessary.

3
Comparisons of 3D Protein Structures from Hyperthermophiles with Their Mesophilic and Thermophilic Counterparts – Surface Saltbridges – A Dominant Theme

To date, structural information is available for, say a dozen, of hyperthermophilic proteins and enzymes that have been isolated from microorganisms, able to grow around 100°C. For some of them the crystal- or NMR structures are also known for the homologous mesophilic or thermophilic proteins and provide ample opportunities for structural comparisons. Rubredoxin from *Pyrococcus furiosus* was the first of the hyperthermophilic proteins to be characterized by X-ray diffraction methods at 1.7-Å resolution (Fig. 4) [42]. The structure of this protein has also been determined independently by NMR spectroscopy [43].

In addition, the secondary structure of ferredoxin from *Pyrococcus furiosus* has been determined, again by NMR [44]. The structures of both pyrococcal proteins turned out to be surprisingly similar to their mesophilic counterparts. They showed identical folding topologies and highly conserved hydrophobic cores. Their increased thermotolerance could be explained by some minor changes involving mostly residues placed on the molecular surface. By these changes some secondary structure elements were modified such as a slightly longer α-helix and a triple- rather than a double-stranded β-sheet. Some specific ionic interactions at N- and C-termini were also modified.

A structural comparison of the 1.75-Å crystal structure of ferredoxin from *Thermotoga maritima* with that of ferredoxins from mesophilic microorganisms has suggested that the extreme thermostability of the Tm ferredoxin is achieved without large changes of the overall structure [45]. A number of

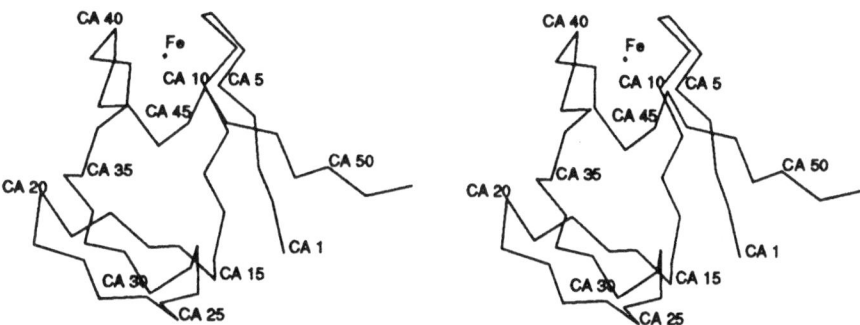

Fig. 4. C^α diagram showing the overal folding of Rubredoxin from *Pyrococcus furiosus*; the iron atom (Fe) is at the top of the molecule; N- and C-terminus are close together (pendant tails *on the right*); figure taken from [42]

potentially stabilizing features has been observed. These include stabilization of helices by extension, replacement of residues in strained conformation by glycines, strong docking of the N-terminus and an overall increase in the number of hydrogen bonds. An increase of the number of saltbridges has not been found. The observed changes have been suggested to lead to decreased flexibility of the protein.

Further insight into the structural determinants of increased protein thermal stability has recently been provided by the crystal structure determinations of three large hyperthermophilic enzymes: glyceraldehyde-3-phosphate dehydrogenase from *Thermotoga maritima* (Tm GAPDH) [46], of Glutamate dehydrogenase from *Pyrococcus furiosus* (Pf GluDH) [47] and Glutamate Dehydrogenase from *Thermotoga maritima* (Tm GluDH) [8]. *Thermotoga maritima* is a hyperthermophilic eubacterium that grows strictly anaerobically at about 55–90 °C. In all of these cases specific comparisons could be made with the crystal structures of homologous enzymes from mesophilic or thermophilic organisms and provided important insights into possible structural determinants of protein thermostability.

The structure of Tm GAPDH is highly homologous to the enzyme from other thermophilic and mesophilic bacteria (Fig. 5a). By comparing the sequences of the enzymes from Tm, *Bacillus stearothermophilus* (Bs), and *Thermus aquaticus* (Ta), 63 and 59% sequence identity have been observed with only 8% of the exchanges of nonconservative nature. By comparing all known structures, it has been shown that around 33% of the residues were identical [48]. The X-ray analysis has yielded a topology which is closely similar to the structure of the enzyme from lobster, (Ha), (by the way, a psychrophile!) and Bs (Fig. 5b).

In particular, the catalytic and the NAD binding domains have shown a close structural similarity by superposition. Fitting both domains separately has resulted in rms deviations of superimposed C^α positions of 0.57 and 0.83 Å, respectively. These figures are rather small. Insertions and deletions have been found exclusively in surface loops of the molecule. The most significant structural difference, that presently is being confirmed in an increasing number of

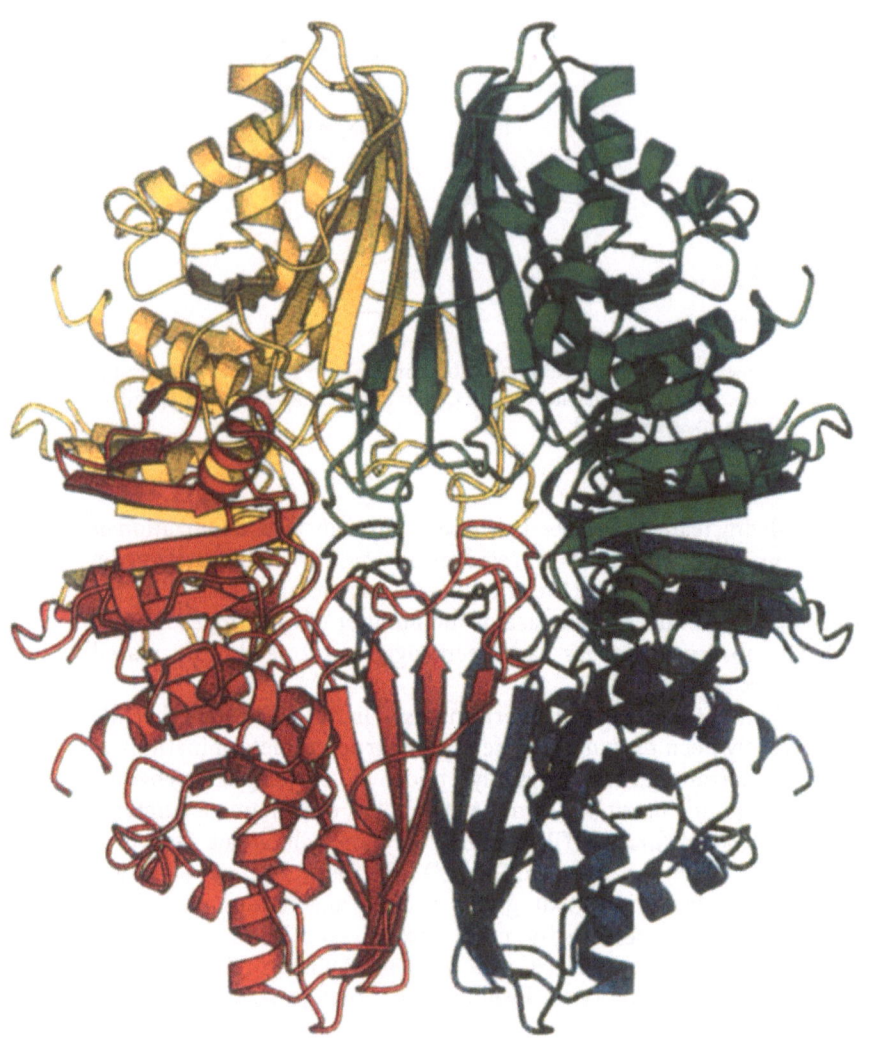

Fig. 5a. Ribbon diagram showing the structure of the GAPDH tetramer from *Thermotoga maritima*; subunits are indicated in *red, green, blue and yellow*

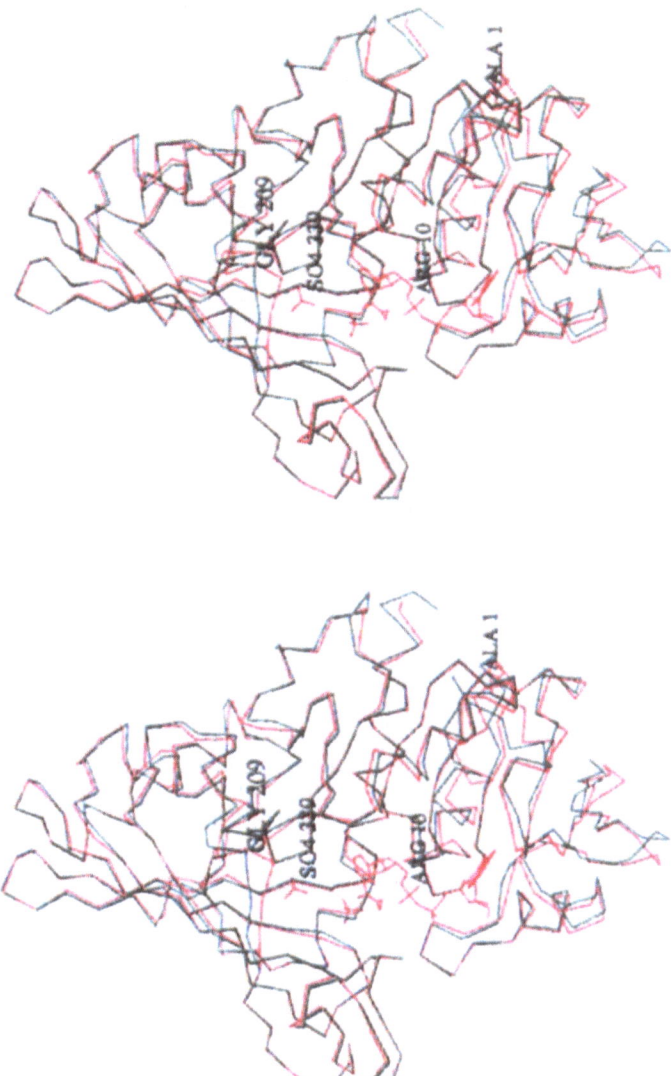

Fig. 5 b. Stereo diagram of the C$^\alpha$ backbones of Tm GAPDH (*red*) and Bs GAPDH (*green*) after least squares superposition of the catalytic domains; different orientations of the NAD$^+$ binding domains relative to the catalytic domains can be recognized (figure taken from [46])

other hyperthermophilic proteins, is a network of charge-charge interactions involved in the formation of surface saltbridges. Specifically, the number of ion pairs (d < 4 Å) that has been found in Ha, Bs and Tm GAPDH was 55, 61 and 78, respectively. All attempts, known to date, to correlate specific charged residues with thermostability by "point" mutations have led into a difficult and complex problem which is due to the high cooperativity of the interactions in a protein molecule (see below). Accessible surface has not correlated in an interpretable way with the thermal stability of Ha, Bs and Tm GAPDH. However, for the three structures compared, the surface area of hydrophobic residues buried in tetramer contacts increases significantly, and has been explained by a stabilization of the tetramer with respect to dissociation of the subunits.

3.1
Even More Saltbridges ...

The ready availability of glutamate dehydrogenases (GluDH) has created considerable interest in this member of the dehydrogenase family as a model system for studies of enzyme catalysis at high temperatures and for the analysis of thermostability in a large oligomeric enzyme. Furthermore, the structure of the mesophilic counterpart from the bacterium *Clostridium symbiosum* (Cs) has been solved and refined by X-ray crystallography at high resolution [49] and serves as a basis for structural comparisons.

GluDHs are ubiquitous enzymes which function at an important branch point between carbon and nitrogen metabolism. They catalyze the oxidative deamination of L-glutamate to 2-oxoglutarate and ammonia with the concomitant reduction of NAD(P)+. The majority of these enzymes are hexameric with 32 symmetry and a molecular weight around 240000 Da. Amino acid sequence comparisons have shown that the hyperthermophilic GluDHs display considerable sequence homology with other members of the family [50]. More specifically, Pf GluDH and *Thermotoga maritima* (Tm) GluDH show 35% and 55% sequence identity with mesophilic Cs GluDH. A sequence alignment based on secondary structure for GluDHs from Cs, Pf and Tm is shown in Fig. 6.

The atomic structures of Pf GluDH [47] and Tm GluDH [8] (Fig. 7a, b) have recently been solved by X-ray diffraction and molecular replacement with Cs GluDH as a search structure. Each subunit of this hexameric enzyme is organized into two domains which are connected by a hinge region and separated by a cleft. The basic elements of the folding pattern of both domains are a β-sheet comprising parallel and antiparallel strands flanked on both sides by α-helices. Domain I is largely formed from residues in the N-terminal portion of the polypeptide chain. The residues from this domain are responsible for all the subunit interactions on the twofold and threefold contact sites of the hexameric assembly. Domain II forms the nucleotide binding site and comprises residues from the C-terminal portion. The folding pattern of domain II is similar to the classical nucleotide binding domain (Rossmann fold) in lactate dehydrogenase [51]. The active site is located in the cleft which undergoes closure upon binding of the coenzyme and the amino acid substrate to bring the coenzyme into the proper orientation for the transfer of the hydride anion [49].

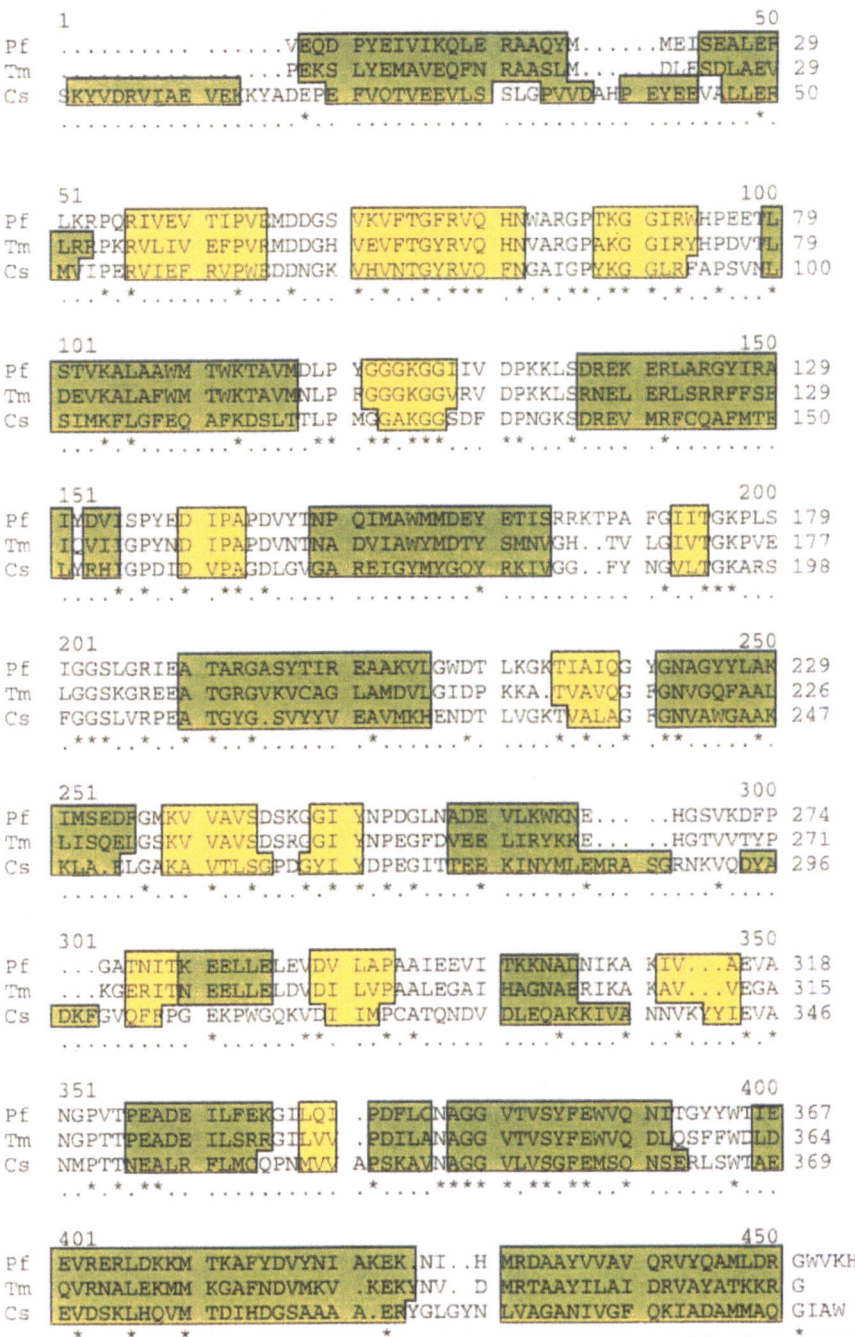

Fig. 6. Structure-based sequence alignment of Pf, Tm and Cs glutamate dehydrogenases assigned with the program PROCHECK (Laskowski et al. (1993) J. Appl. Crystallog. 26:283); α helical regions in *green*, β strand regions *yellow*; (figure taken from [8])

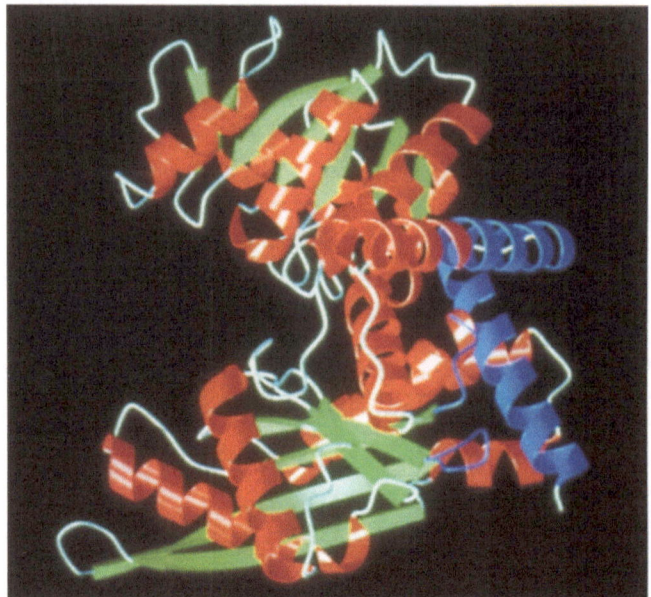

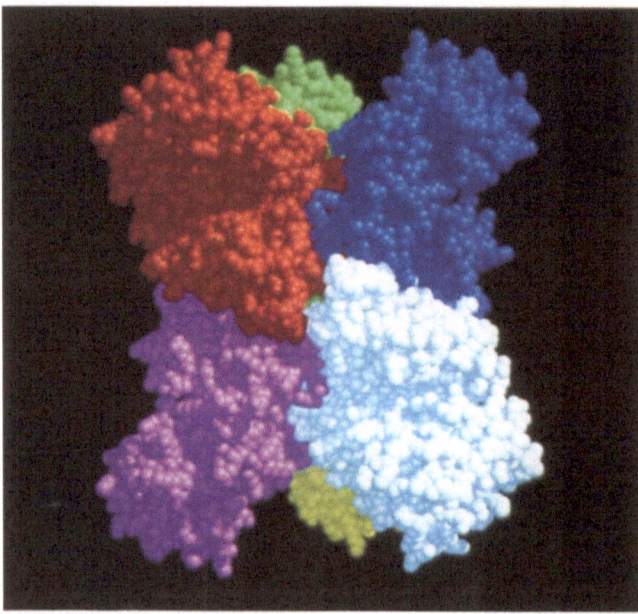

Fig. 7a–b. **a** Main chain folding of a single subunit of Tm glutamate dehydrogenase; α helical regions *red*; β strands *green*; the subunit is organized into two domains, connected by a hinge region; lower domain (I) involved in intersubunit contacts; upper domain (II) contains the NAD(P) binding site. **b** Space-filling representation of a Tm glutamate dehydrogenase hexamer, viewed along a local twofold axis; the hexamer has 32 symmetry; single subunits are shown in different colors

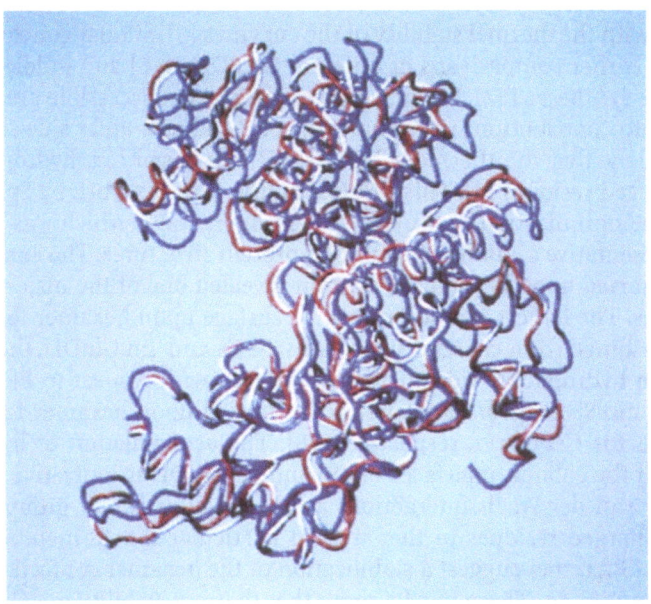

Fig. 7 c. Ribbon drawing of the compared glutamate dehydrogenase structures after least squares superposition; Cs GluDH, *blue*; Tm GluDH, *white*; Pf GluDH, *red*; (figures 7a, c taken from [8])

The folding patterns of both hyperthermophilic GluDHs are quite similar to that of the hexameric Cs GluDH. Ribbon drawings of the subunit structures of Cs, Pf, Tm GluDH are displayed in Fig. 7c, showing similar folding patterns of the three members of this enzyme family. The C^α positions of domain I in Tm differed with an r.m.s.d of 2.2 and 3.2 Å from Pf and Cs, respectively; the analogous figures for domain II (nucleotide binding) were 3.5 and 4.3 Å. The active site regions showed high structural conservation: The C^α positions in Tm GluDH differed only by an r.m.s.d of 0.3 Å and 0.7 Å from the same positions in Pf and Cs, respectively. All of the residues involved in catalysis [48], namely D144, K104, G70, K68, K92, N341, S348 in Tm GluDH, were found to be conserved in the three enzymes.

Detailed comparisons of the structures of the three members of the GluDH family have been possible and have revealed some insights into structural determinants which suggest extreme stability for these enzymes. From the superposition in Fig. 7 c, it is obvious that some of the surface loops appear shortened and more compact in the hyperthermostable members. The structure based sequence alignment, shown in Fig. 6, suggested that an increased thermal stability of the hyperthermophilic members may be correlated to 1) an increase in structure rigidity by a reduction of glycine residues 2) an improvement of hydrophobic contacts in the core of Pf GluDH by valine to isoleucine mutations 3) a modulation of the subunit flexibility via domain movements [8]. The normalized (per residue) accessible surface areas of Cs, Tm and Pf GluDH did not correlate with the thermal stability of the enzymes [8]. Similar conclusions were drawn by earlier comparisons of different GAPDHs [46] and indole-3-glycerol phosphate synthases [52]. When, however, the solvent accessible areas were separated into contributions from charged, polar and non-polar side chains it became evident that the thermostable members exposed less hydrophobic and more charged residues [8]. This result is in line with the work by Spassov et al. [23] on the optimization of protein-solvent interactions which was performed on a representative set of high resolution protein structures. The analysis of the surfaces buried upon hexamer formation revealed one of the major structural differences. The largest fraction of buried surface upon hexamer formation in Pf is contributed from charged side chains. In Cs and Tm GluDH, the contributions from hydrophobic side chains ar favoured with respect to Pf GluDH. In fact, the ratio %charged/%apolar residues buried upon hexamer formation is 1.0:1.1:1.8 for Cs:Tm:Pf, respectively [8]. The accumulation of hydrophobic residues at the subunit interfaces of Tm and Cs GluDH is indicative of a stabilization by van der Waals interactions and removal of apolar groups from the solvent. Charged residues in this area, in particular the 18-member ion-pair network in Pf, rather suggest a stabilization of the hexamer contacts by charge-charge interactions. These results show, that thermal stability may be achieved by different ways, even in members of the same protein family. Thermostable enzymes like Tm GluDH, which function optimally between 60 and 80 °C (in vitro), very likely have a different balance of stabilizing forces than those, like Pf, operating around 100 °C. Similar observations have been reported in a recent structure comparison of GAPDHs [46].

However, the most striking difference revealed by structural comparison was a dramatic increase in the number of ion pairs, which correlated well with the growth temperature of the organisms, the melting temperatures of the proteins and the temperatures of maximum enzymatic activity. Cs GluDH has a melting temperature slightly above 55 °C [53], whereas Tm GluDH and Pf GluDH melt at $T_m = 95$ °C and $T_m = 105$ °C, respectively [54]. The number of ion pairs per hexamer in the hyperthermophilic and mesophilic enzymes is shown in Table 3.

Whereas the percentage of charged residues forming ion pairs is similar in Cs and Tm GluDH, there is a significant increase in Pf GluDH. Furthermore there is a similar preference for the utilization of arginine side-chains in the formation of ion pairs in both hyperthermophilic proteins. However, the increased involvement of arginine side chains in the formation of ion pair interactions is not due to the percentage of such residues in the sequences of both hyperthermophilic enzymes but is rather due to the role which those side chains have in the structural context. The preference for arginine side chains may result from the ability of arginine to participate in multidentate interactions.

Many of the ion pairs in the hyperthermophilic members are organized in networks, which are found on the surface or buried at domain and subunit interfaces, whereas in the mesophilic Cs GluDH the majority of ion pairs involve an isolated linkage of one residue to another. The tendency that ion pairs are organized in extensive interacting networks is most pronounced in Pf GluDH

Table 3. Ion pair statistics for bacterial and archaebacterial Glutamate dehydrogenases

	Cs	Tm	Pf
Melting temperature (°C)	55	93	113
No. ion pairs per hexamer	188 (308)	223 (362)	288 (462)
No. ion pairs per residue	0.07 (0.11)	0.09 (0.15)	0.11 (0.18)
% of charged residues forming ion pairs	48 (70)	49 (73)	58 (82)
% of ion pairs formed by Arg/Lys/His	47/33/20	58/35/7	61/35/4
% of ion pairs formed by Glu/Asp	56/44	58/42	53/47
% of all Arg forming ion pairs	61 (78)	68 (88)	90 (100)
No. residues forming two ion pairs	40 (83)	78 (98)	118 (210)
No. residues forming three ion pairs	10 (34)	18 (49)	24 (74)
No. 2/3/4 member networks	72/24/12	66/33/18	54/24/12
No. 5/6/7/18 member networks	0/0/0/0	0/0/6/0	12/6/0/3
No. of intersubunit ion pairs	36 (54)	34 (73)	54 (90)

Ion pairs were evaluated with a distance of less than or equal 4 Å (or 6 Å for the numbers in parentheses) between charged groups. Table taken from [8].

(Fig. 8 a) [47]. There is a clear trend for ion pair network formation also in the hyperthermophilic Tm GluDH, however in this member the formation of smaller 3 or 4 member networks seems to be predominant. The largest network in Tm GluDH involves 7 members and is found in the cleft between domain I and II, (Fig. 8 b) [8]. In Pf GluDH the most extended ion pair cluster lies at the interface between dimers and involves twofold symmetric interactions of 18 charged residues. In addition, several networks with 5 and 6 members have been detected.

An increase in the number of intersubunit ion pairs reported for the Pf enzyme [47] was not found in Tm GluDH in which the number of intersubunit ion pairs is even smaller compared to the Cs enzyme.

From structural studies as well as from recent work on the optimization of protein-solvent interactions [17, 23] in thermostable and hyperthermostable proteins evidence has accumulated for a statistical increase of the observed number of ion pairs which may suggest a possible role in the formation of stabilizing interactions. The structure studies include work on indole-3-glycerol phosphate synthase from *Sulfolobus solfataricus* [52], aldehyde-ferredoxin oxidoreductase [27] and glutamate dehydrogenases from *Pyrococcus furiosus* [47] and *Thermotoga maritima* [8], glyceraldehyde-3-phosphate dehydrogenase from *Thermotoga maritima* [46], ribonuclease from *Thermus thermophilus* [55] and malate dehydrogenase from *Thermus flavus* [56]. The analysis of these structures has shown that multidentate ion pair interactions are quite frequent but less extended than those observed in Pf and Tm GluDH. In conclusion, the results of X-ray structure analyses point out a clear trend: ion pair interactions are both more dominant and more complex in hyperthermostable enzymes.

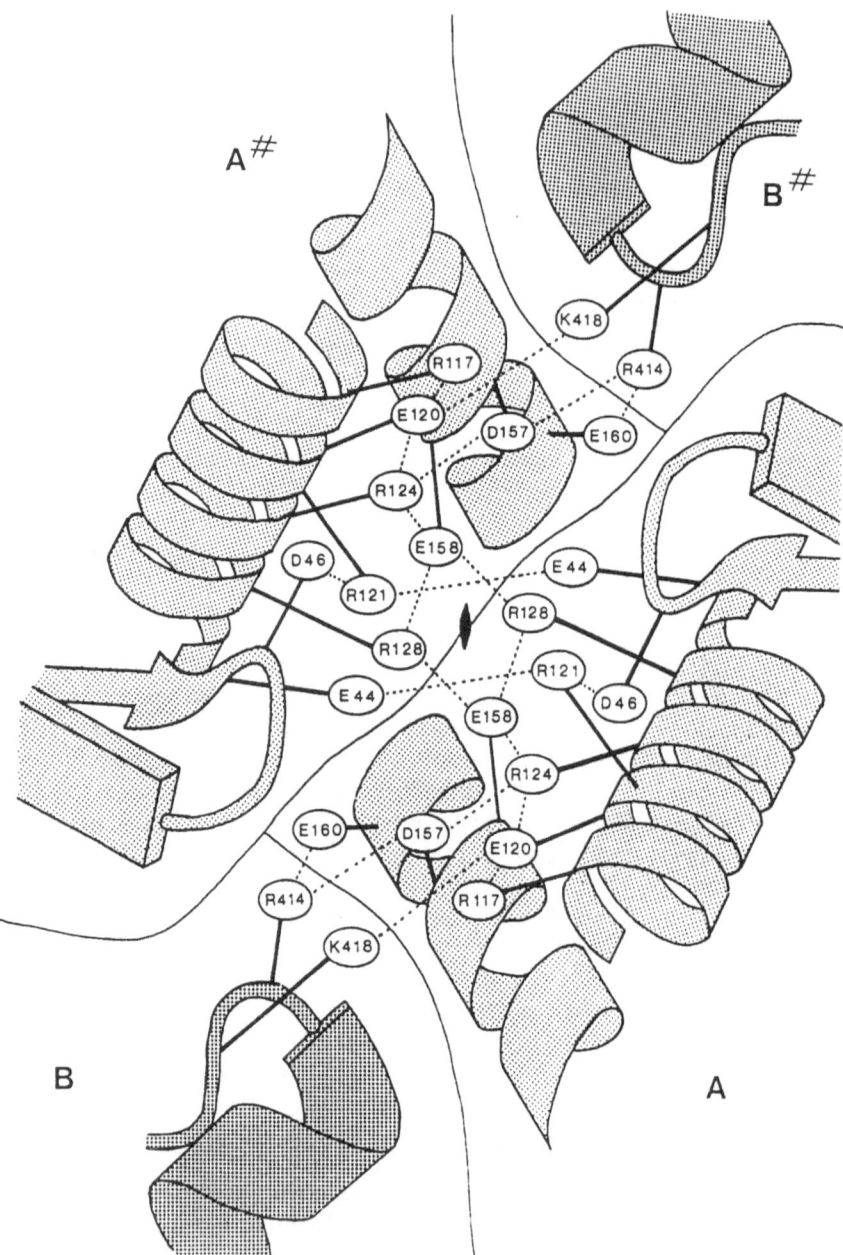

Fig. 8a. Schematic representation of the ion pair network at one of the twofold axes relating dimers (●) of the Pf GluDH hexamer; two of the subunits of each trimer are labelled *A* and *B* with their respective twofold-symmetry related partners *A#*, *B#*; Individual amino acids involved in ion pair interactions are indicated using the one-letter amino acid code, followed by their sequence number; (figure taken from [47])

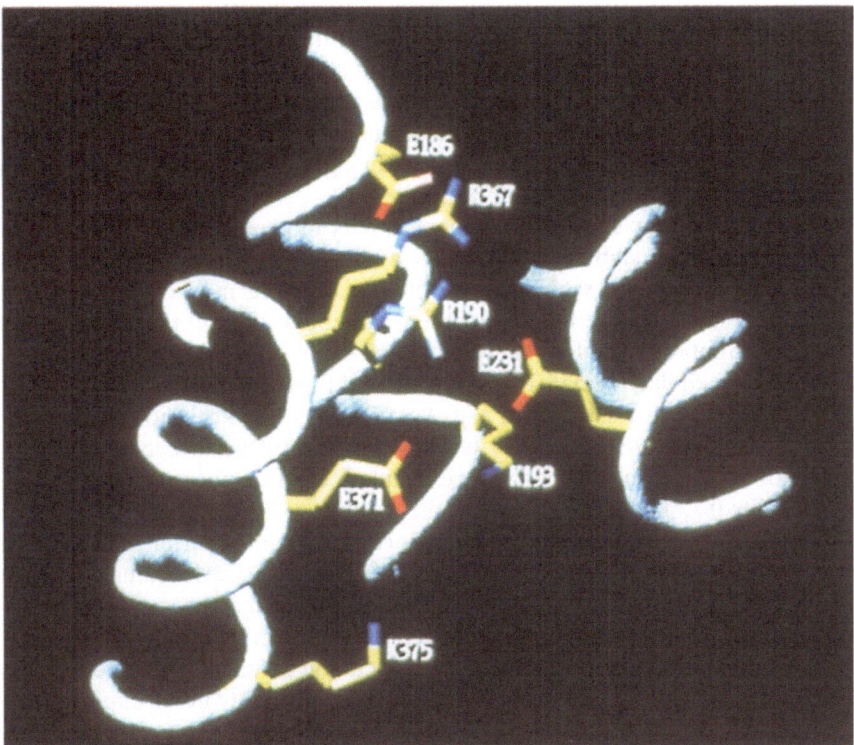

Fig. 8b. Largest seven residue ion pair network in Tm glutamate dehydrogenase; The network is found in a cleft at the hinge regions between domain I and domain II; The network links three α helical structure elements; (figure taken from [8])

3.2
Salt Bridges and Salt Bridge Networks on the Protein Surface – Major Determinants of Thermal Stability?

In 1975, Perutz and colleagues have published a paper on the heat stability of bacterial ferredoxins and hemoglobin A2 in Nature [57]. They wrote: "Most enzymes are quickly inactivated above about 55 °C, but those from thermophile bacteria are stable for long periods at higher temperatures. We do not know why, because so far their structures have proved too complex ..." At the end of the same paper they suggested: "... it seems that in monomeric enzymes the extra energy of stabilization can be provided without disturbance of the tertiary structure by a few extra salt bridges on the molecular surface; in oligomeric enzymes by some extra salt bridges, hydrogen bonds or non-polar bonds at the subunit interfaces".

Their vision is more than up to date in this research field today and has received much support from recent high resolution structure studies on hyperthermostable enzymes: but do we now, 20 years later, really "know why" the surface saltbridges are there? They may also be there because the structures have adapted to still unknown biological or environmental constraints which have nothing to do with extreme temperature and stability. Cooperativity of surface salt bridges could be of importance to molecular mechanisms of allosteric regulation. There are a number of examples for proteins where in the transition from one allosteric state to the other, salt bridges are broken with the missing contacts being replaced by water molecules.

In a paper on the strength and cooperativity of contributions of surface salt bridges to protein stability by Horovitz et al. [58] a triad of charged residues that form two exposed saltbridges on the surface of barnase, comprising Asp8, Asp12 and Arg110, have been mutated to alanine to give all the single, double and triple mutants. The free energies of unfolding of wild type and seven mutants have been determined. In the intact triad the contribution to stabilization of the protein of the salt bridge Asp12–Arg110 is –1.25 kcal/mol, whereas that of the salt bridge Asp8–Arg110 is –0.98 kcal/mol. Due to cooperativity, the strengths of the two salt bridges are coupled: the free energy of each pair is reduced by 0.77 kcal/mol when the other is absent. Furthermore it was found, that the salt-linked triad, relative to alanines at the same position, does not contribute to the stability of barnase since the favourable contributions of the saltbridges are offset by other electrostatic and non-electrostatic energy terms. The interesting and surprising result, however, is the quite high coupling free energy between the pairs, which more or less reflects the actual cooperative situation in an ion pair network. This contribution could certainly be of importance to the stability and function of the protein.

The side chains of Arg31, Glu36 and Arg40 in the Arc repressor form a buried saltbridge triad [59]. The small network has been replaced by hydrophobic residues in combinatorial mutation experiments resulting in biologically active mutants that are thermodynamically more stable than the wild type protein. Thus, simple hydrophobic interactions provide more stabilization energy than the buried saltbridge network. However, some observations made by these authors have indicated cooperativity defects (DNA binding) in the mutants.

No satisfactory explanation based on experimental data from hyperthermostable proteins is available to date which could describe how surface ion pairs or extended ion pair networks might contribute to the stabilization energy. This explanation must come from site-directed mutagenesis experiments in combination with the measurement of thermodynamic stability quantities on the mutant structures. Measurements of activity parameters and melting temperatures alone cannot provide direct proof that the respective mutation of an ion pair is coupled to (thermodynamic) stability. To start with, the destabilization of a protein by the removal of key interactions seems to be the most feasible and logical approach.

3.3
Single Point Mutations, Pairwise Mutations ... Increasing Complications

Mutation studies with the goal of recognizing the structure elements which determine thermostability in a large protein seem far from being straightforward, since interactions in protein molecules are highly cooperative. Thus the effect of a single mutation is often not only local, and structural differences between mutants are usually not strictly focused at the exchanged residues but spread over large parts of the molecule. Often function as well as stability are influenced at the same time.

An experimental approach to estimate the strength of an interaction is to mutate one of the residues involved in the interaction and measure the change in stability of the protein. Such single mutation experiments can be difficult to interpret since, in many cases, not only the interaction of interest is disrupted but other interactions, both direct and indirect, as well. An approach applicable to pairwise interactions is to invoke a double-mutant cycle [60, 61], which consists of wild-type protein, the two single mutants and the double mutant where both residues are replaced. The effects of interactions, other than those of the pair, often cancel out, thus allowing the quantification of an interaction between two residues. However, any single cycle provides no information of a third position in the interaction. Constructing multiple thermodynamic cycles enables the measurement of all three pairwise interactions and also the determination of the (important!) energetic coupling between these interactions [58, 61].

Further complications arise from the experimental difficulties of determining thermodynamic quantities, like the free energy of stabilization ΔG, for large oligomeric enzymes. Such systems often do not unfold in a reversible two-state process [62] or show a tendency to denature irreversibly in thermal unfolding mesurements using differential scanning microcalorimetry. There is a possibility of circumventing this difficulty through monitoring equilibrium unfolding of the protein by denaturation in urea using circular dichroism- or fluorescence spectroscopy. It has been shown empirically, that the free energy of unfolding ΔG^u is linearly related to the urea concentration [63]. However, the long extrapolation from the transition region, usually from around 4 M urea to 0 M urea, can lead to large errors in the estimates of the unfolding free energy in water, $\Delta G^u (H_2O)$.

4
Excursion into Protein Thermodynamics: Stability, Determined by Thermal Unfolding of Small Proteins from Hyperthermophiles

Small single domain proteins or isolated domains of larger proteins may serve as model systems for multidomain proteins which, due to the complexity and irreversibility of their unfolding transitions cannot be analyzed by equilibrium thermodynamics.

Several characteristics of Sso7d, a small 7 kDa DNA binding protein from *Sulfolobus solfataricus*, make it an attractive model for protein stability studies at high temperatures: Sso7d is a single domain protein which does not aggregate even at high protein concentration, it contains no metal binding sites and disulphide bridges, furthermore its 3D structure is known [64].

The Sso7d fold comprises a triple-stranded β-sheet onto which a double-stranded sheet is packed in almost rectangular fashion and a small C-terminal two turn α-helix. This structure motif is surprisingly similar in its topology to eukaryotic Src-homology 3 (SH3) domains. A search in the PDB for similar 3D structures and topology [65] revealed the conservation of this folding pattern in all three domains of life: Eukarya, Archaea and Bacteria. This folding motif may have evolved at high temperature early in evolution, and has probably conserved its thermotolerance.

The tertiary structure of Sac7d, a similar protein from *Sulfolobus acidocaldarius*, shows the same folding pattern as Sso7d, except for a C-terminal extension of the α-helix (Fig. 9) [66].

Sso7d belongs to a family of small basic proteins which play a role in the structural organization of DNA into a chromatin-like structure. In in vitro experiments they increase, upon non-specific binding, the melting temperature of DNA by about 40 °C [64]. These proteins are called "histone-like" and share physical properties with eukaryotic histones but show no sequence homology [67].

The energetics of thermal unfolding of Sso7d has been studied by differential scanning microcalorimetry and circular dichroism spectroscopy [68]. This work represents the first quantitative thermodynamic characterization of a protein from a hyperthermophilic archaebacterium. Preliminary calorimetric studies have been reported for Glutamate dehydrogenase from *Pyrococcus furiosus* [69] and for the Sac7d protein from *Sulfolobus acidocaldarius* [70].

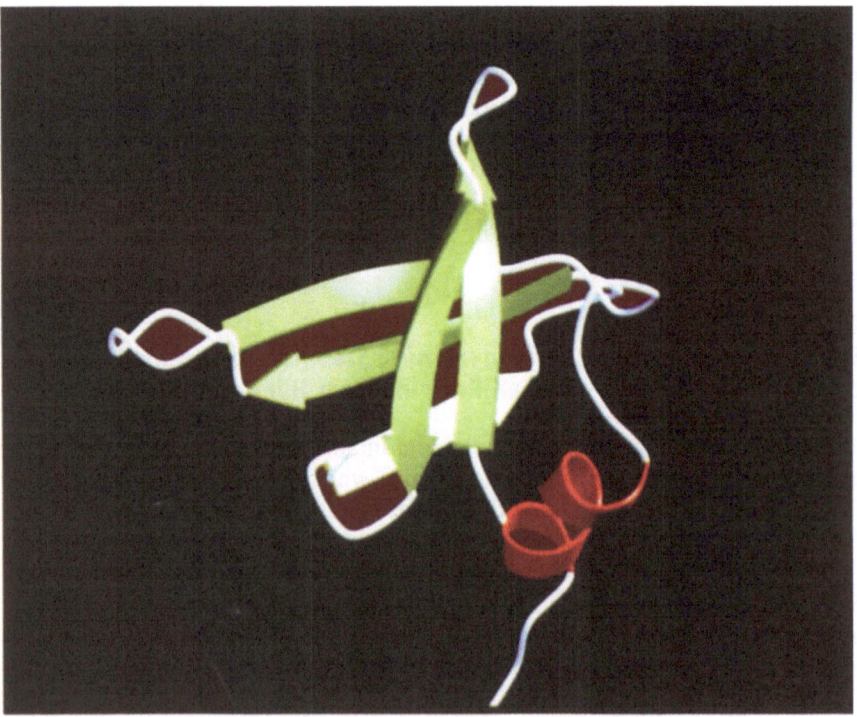

Fig. 9. Ribbon diagram showing the folding pattern of the DNA binding protein Sso7d from *Sulfolobus solfataricus*; α-helix *red*; β-strands *green*; loop regions *white*

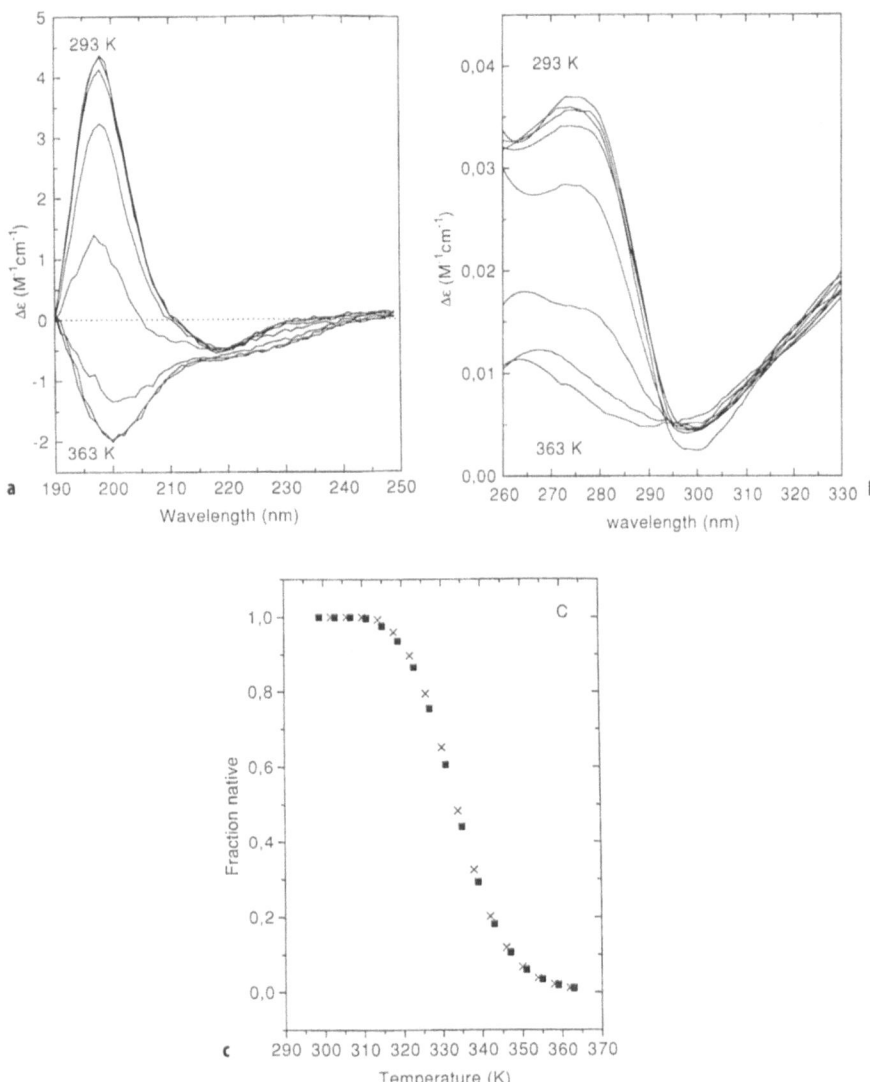

Fig. 10a–c. Thermal denaturation of Sso7d measured by CD spectroscopy at different wavelengths (pH 2.5, 5 mM sodium phosphate buffer); The molar circular dichroism $\Delta\varepsilon$ (M^{-1} cm^{-1}) per peptide bond is displayed; **a** region of CD spectra reporting changes in the secondary structure ($\lambda = 190–250$ nm); protein concentration 0.2 mg/ml; **b** region of CD spectra indicative for changes in the environment of aromatic residues ($\lambda = 260–330$ nm) thus representing changes in tertiary structure; protein concentration 2.5 mg/ml; **c** Normalized fitted function calculated from CD data measured at $\lambda = 200$ nm (×) and at $\lambda = 275$ nm (■); (figure taken from [68])

Protein unfolding is conveniently described by a two-state process: (i) melting of the solid-like core into a liquid-like state; (ii) hydration of nonpolar groups, following the desorganization of the liquid-like state.

For small globular proteins with a reversible two state unfolding transition the change in free energy between the native (N) and the denatured state (D) is given by

$$\Delta G = G^D - G^N = \Delta H - T\Delta S \tag{6}$$

The free energy of unfolding ΔG can be obtained from Eq. (6) as a function of temperature when the temperature dependence of ΔH and ΔS is known. In order to calculate the enthalpy ΔH and entropy term ΔS, the difference in heat capacity between the native and denatured state, ΔC_p, must be measured. This quantity can be determined directly from differential scanning calorimetric measurements (see Fig. 11) or more indirectly by spectroscopic techniques.

Assuming that ΔC_p is constant in the relevant temperature range the temperature dependence of the enthalpy and entropy term is given by

$$\Delta H(T) = \Delta H(T_m) + (T - T_m)\Delta C_p \tag{7}$$

$$\Delta S(T) = \Delta H(T_m)/T_m + \Delta C_p \ln T/T_m \tag{8}$$

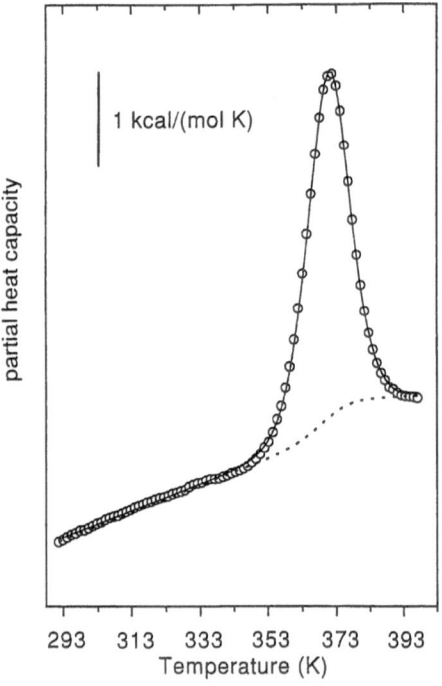

Fig. 11. Representative differential scanning calorimetric measurement on recombinant Sso7d in 5 mM sodium phosphate buffer, pH 5.0; The *continuous line* shows the fit to a two-state model N ⇔ D, using software supplied by MICROCAL; Every third data point (O) is displayed; the progress baseline is shown as a *dotted line*; (figure taken from [68])

where T_m represents the transition (i.e. melting) temperature. The stability curve can finally be calculated from the quantities above using a modified Gibbs-Helmholtz equation [70]:

$$\Delta G(T) = \Delta H(T_m) - T \Delta S(T_m) + \Delta C_p((T - T_m) - T\ln T/T_m) \quad\quad (9)$$

$\Delta G(T)$ is a direct measure of the conformational stability at any temperature and is suitable to describe quantitatively the thermodynamic stability of a protein.

The function $\Delta G(T)$ intersects the x-axis twice ($\Delta G(T_m) = 0$) indicating that proteins denature at high (heat denaturation at T_m) as well as low temperatures (cold denaturation at T'_m).The stability curve of a protein is determined by its slope $\partial\Delta G/\partial T = -\Delta S$, by the curvature $\partial^2\Delta G/\partial T^2 = -\Delta C_p/T$ and the position of the ΔG maximum, where $\Delta S = 0$ and stabilization at the corresponding temperature is only due to enthalpic contributions [71].

The heat capacity difference ΔC_p between the native and thermally denatured state of Sso7d has been determined directly from calorimetric measurements and from the linear ΔH_m vs T_m dependence obtained from CD measurements [68].

A value of 620 cal mol^{-1} K^{-1} has been obtained, which corresponds to a specific ΔC_p, calculated per mol of amino acid residue, of 9.7 cal mol^{-1} K^{-1} residue^{-1}. This value is at the lower end of the values found for globular proteins [72, 73]. The main contribution ($\approx 95\%$) to the heat capacity change is due to the hydration of non-polar residues which are exposed upon unfolding and to a much lesser extent ($\approx 5\%$) by the loss of non-covalent interactions. The heat capacity change is directly proportional to the exposed non-polar surface area [74] upon unfolding. This function can be written in terms of the polar and non-polar accessible surface areas and the total area buried from the solvent [75]. The small heat capacity change of Sso7d may result from the fact that it is a quite small protein unable to form a large hydrophobic core. Furthermore, a considerable proportion of hydrophobic residues are already exposed in the native state. Another explanation would be an only partial exposure of hydrophobic residues in the unfolded state.

The temperature dependence of the free energy of unfolding $\Delta G(T)$ is completely specified by the three parameters T_m, ΔH_m and ΔC_p or equivalently by ΔH_m, ΔS_m and ΔC_p (see Eq (9)). Sso7d is a protein with a particularly high melting temperature ($T_m = 98.9\,°C$ at pH 5.5) (Fig. 12), therefore there is a considerable interest, how the thermostability of this protein is reflected by the thermodynamic parameters.

The melting temperature, T_m, of a protein can be increased by variations in ΔH_m, ΔS_m and ΔC_p that result in a certain combination of lifting or shifting the stability curve and/or flattening its curvature, i.e. increasing its breadth [76]. Within the framework of an approximative treatment [77] the native form of a protein is stable over a temperature range $\Delta T \approx 2(\Delta H_m/\Delta C_p)$ and the melting temperature is connected to the temperature of maximum stability by $T_m \approx T^* + (\Delta H_m/\Delta C_p)$.

Sso7d has its maximum stability at a temperature of 9 °C with a free stabilization energy of $\Delta G_{max} \approx 7$ kcal mol^{-1}. At the temperature of optimal growth

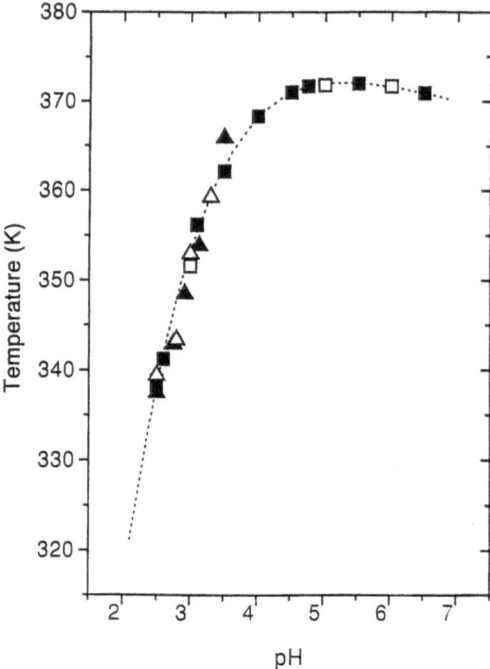

Fig. 12. pH dependence of the melting temperature (T_m) of Sso7d from *Sulfolobus solfataricus* calculated from data of recombinant (DSC(■), CD(□)) and native (DSC(▲), CD(△)); (figure taken from [68])

($\approx 80\,°C$) of *Sulfolobus solfataricus*, the protein is only moderately stabilized with a $\Delta G \approx 2.8$ kcal mol^{-1} and at the maximum growth temperature ($\approx 90\,°C$) ΔG drops to about 1 kcal mol^{-1}. Sso7d does certainly not show a particularly stable native conformation at physiological growth temperatures of the host. However, thermal unfolding in vitro may be different from the situation in vivo where the protein may be externally stabilized by interaction with DNA and/or intercellular solutes. The specific denaturation enthalpy $\Delta H\,(25\,°C)$ of 0.28 kcal mol^{-1} residue^{-1} and the denaturation entropy $\Delta S = 0.5$ cal mol^{-1} K^{-1} residue^{-1} are in the usual range for small proteins [78]. At the convergence temperatures [79], $T_h{}^* = 377$ K, $T_s{}^* = 385$ K, significantly lower specific enthalpy and entropy values, $\Delta H^* = 1.04$ kcal mol^{-1} residue^{-1} and $\Delta S^* = 2.99$ cal mol^{-1} K^{-1} residue^{-1} were found for Sso7d. According to Ragone and Colonna [80] both parameters constitute a residual unfolding free energy, $\Delta G^* = \Delta H^* - T\Delta S^*$, which is assumed to be devoid of the hydration contribution, thus representing the energetics associated only with van der Waals and hydrogen bond interactions. It has been suggested that the associated temperature, defined by $\Delta H^*/\Delta S^*$, reflects the pure melting of the solid-like protein core. Simulations of stability curves have shown that decreasing values of both ΔH^* and ΔS^* shift the unfolding free energy curve toward higher temperatures, and thus higher T_m [80].

The thermal unfolding studies on Sso7d have shown that T_m is not synonymous with larger maximum stability ΔG_{max}, contradicting the often discussed

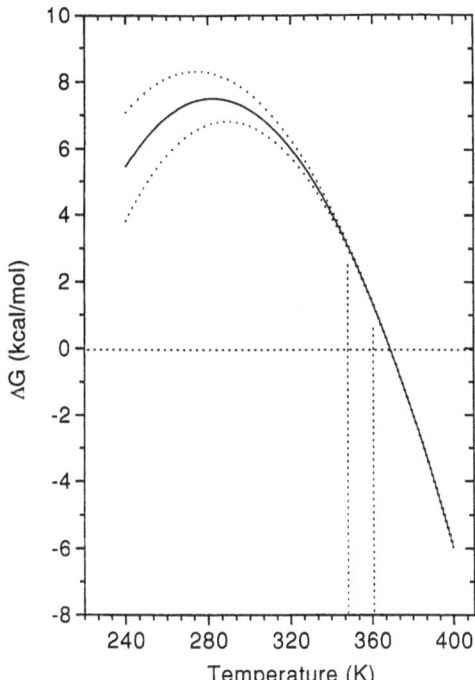

Fig. 13. Stability curve ($\Delta G = f(T)$) calculated under conditions of maximum stability of Sso7d; a heat capacity change of 620 cal mol^{-1} K^{-1} was used; *continuous line*: unfolding free energy of Sso7d calculated from equation (9); *dotted lines*: indication of the error of the measurement ($\pm$ 10%); *vertical dotted lines*: optimum and maximum growth temperature of *Sulfolobus solfataricus*; (figure taken from [68])

opinion, that an increased melting temperature T_m correlates with high stability (Fig. 13). The reasons for the high melting temperature of Sso7d are a shallow stability curve with a broad ΔG maximum, corresponding to the small ΔCp value for the unfolding transition. Consequently, proteins of limited thermodynamic stabilization may show an increased melting temperature T_m and thermal stability (T_m) can paradoxically be increased on cost of thermodynamic stability (ΔG and ΔG^*). An explanation in structural terms has been given by assuming a weakening of van der Waals and hydrogen bond interactions [80]. Proteins with significantly higher convergence enthalpies and entropies, like barnase [81] and RNAse T1 [82] show relatively large unfolding free energies yet relatively low melting temperatures.

In the work of McAfee et al. [70] a significant difference in the thermostabilities of native and recombinant Sac7d has been reported ($\Delta T_m = 6.5\,^{\circ}$C). The only known structural difference between both proteins is the amino monomethylation of lysines 5 and 7 in the native protein and the initiating methionine in the recombinant protein. It has been claimed that *Sulfolobus* can increase the thermostability of some of its proteins by specific lysine monomethylation. The calorimetric measurements on methylated and non-methylated Sso7d did not, however, provide evidence for lysine methylation as a mechanism which is used by *Sulfolobus* to increase protein thermostability [68].

5
Extrinsic Stabilization and Acquisition of Thermotolerance –
Chaperonins Catalyze Folding and Repair of Damaged Proteins,
at Least in Vitro

The high protein concentration in a living cell is known to represent an important stabilizing factor for an individual protein molecule. There is, furthermore, a large number of different solutes present (some in high concentration) which can specifically influence the stability of proteins: 1) substrates and cofactors often stabilize the native state of an enzyme or an intermediate on the folding pathway [84], 2) so-called compatible solutes are excluded from the protein surface by specific mechanisms and influence protein stability indirectly by alterations of the hydration shell [85, 86], 3) molecular chaperones, which are, in bacteria and archaea, often synthesized in increased amounts under heat shock conditions, facilitate the correct folding of a polypeptide chain and prevent the aggregation of unfolded proteins [87, 88]. Chaperones are essential for the acquisition of thermotolerance [89] but are also required under normal physiological conditions. In the following, the known archaebacterial chaperonins will be discussed in more detail.

Based on their primary structures, chaperonins can be divided into two families: The GroEL family found in eubacteria and mitochondria and the TCP1 protein family found in archaebacteria and the eucaryotic cytosol [90]. Chaperonins from archaea and members from the TCP-1 protein family are only distantly related to the bacterial GroEL type chaperonins. However, their molecular appearance is similar, except the symmetry which seems to be generally seven-fold in the bacterial chaperonins. Chaperonins from Eucarya and Archaebacteria have the same cage-like toroidal structures [91–94] which are usually built up from 55–60 kDa subunits and show 6-fold, 8-fold or 9-fold symmetry. In some cases two different types of subunits have been described [91, 92, 95–97]. The first member of the TCP1 family characterized in archaebacteria was the temperature factor 55 (TF55) from *Sulfolobus shibatae* [98]. Binding and refolding of denatured proteins have been demonstrated for TF55 in vitro and its ATPase activity at the optimal growth temperature of the organism (75°C) has been found to be similar to that of the well studied chaperonin GroEL from *E. coli* [98, 99]. TF55 is the major heat shock protein observed after transferring *Sulfolobus* species from its optimal growth temperature to the near lethal temperature of 88°C; consequently, large amounts of the chaperonin could be isolated from heat-shocked *Sulfolobus solfataricus* cells [97]. The synthesis of this chaperonin is probably correlated with the thermotolerance of the organism [89].

The structure of TF55 from the archaebacterium *Pyrodictium occultum* has been intensively studied by electron microscopy of single particles and 3D image reconstruction (Fig. 14) [91, 92]. This chaperonin forms a toroidal hexadecameric complex of two eight subunit rings consisting of two different poly-

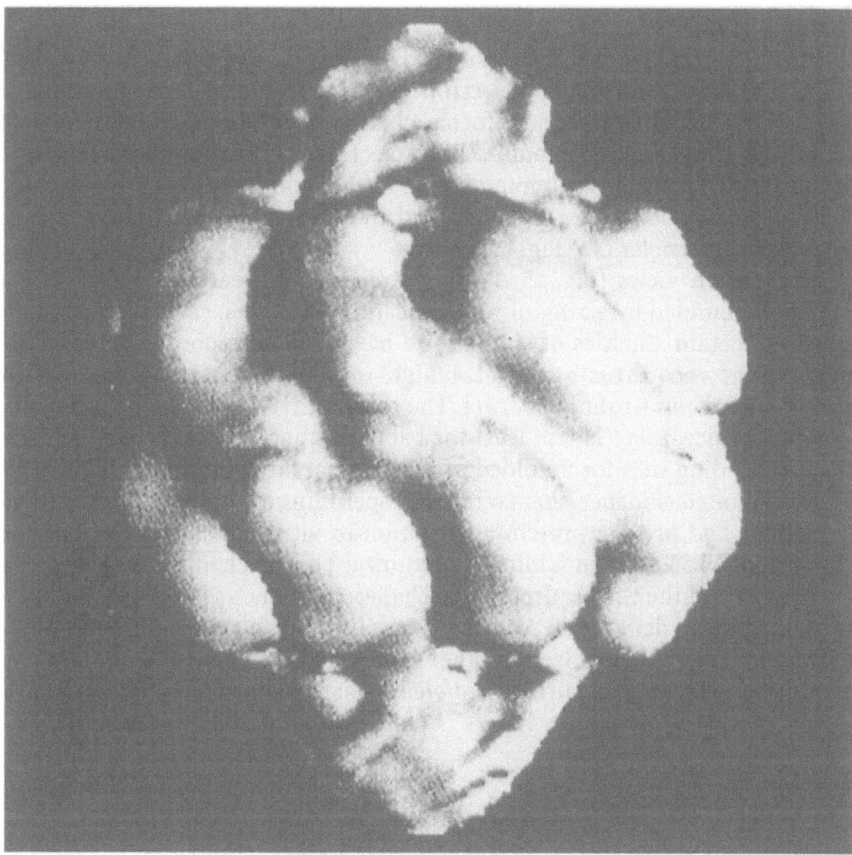

Fig. 14. Three-dimensional model of the thermosome from *Pyrodictium occultum*; Surface-shaded solid model viewed along a twofold axis; two rings containing eight kidney-shaped subunits jeach are stacked such that equivalent faces are in contact; the rings enclose a central cavity which is partially capped at both ends by a mass with narrow connections to the rings; height 190 Å, diameter 165 Å; (figure taken from [92])

peptides with molecular masses of 56 and 59 kDa in a 1:1 ratio. The subunits appear asymmetric and oriented identically in the ring. Two striations, seen in the side views, correspond to each ring suggesting that the subunits consist of at least two domains. The reconstruction shows a large central cavity with a mean diameter of 67 Å. The opening of the cavity on each face is partially occluded by a solid mass which appears to be joined to the ring by narrow extensions. These features at the cavity openings have been explained in three different ways: 1) the mass may be formed by coalescence of domains extending up and towards the centre of the ring from every subunit, or from every second subunit, 2) alternatively, it might be an equivalent to GroES which binds as a seven-membered ring over one mouth of the GroEL tetradecamer, 3) a third and appealing possibility is that the GroEL/GroES equivalents are integrated into a single

polypeptide with the GroES-related domains in a position over the mouth of
the cavity [92].

For TF55 from *Sulfolobus shibatae* and *Sulfolobus solfataricus* nine-fold symmetry has been confirmed by electron microscopy and 2D image reconstruction [93, 97, 98]. The chaperonin forms a hetero-oligomeric complex of two different, but closely related subunits. 2D projections of the chaperonin have been
reconstructed from electron microscopic images and show a nine-fold symme-
trical complex about 175 Å in height and 160 Å in diameter, with a central cavity of 45 Å diameter (see Fig. 15)

The end-on views (Fig. 15 A) show a circular appearance with a central
cavity surrounded by a ring of nine repeating units. The cavity appears to be
filled with stain. Cavities of similar size have also been observed in electron
microscopic reconstructions and a high resolution X-ray structure of the
E. coli chaperonin GroEL [100, 101]. The central cavity seems to be of functional importance; there is – at least for bacterial chaperonins – some evidence
that the binding sites for unfolded proteins are in the central cavity [102, 103].
There is limited evidence, that archaeal chaperonins too assist in folding of polypeptides and prevent protein aggregation in vitro [104]. However, the mechanism of TF55 assisted folding is unknown. This mechanism is likely to differ from that of the GroEL/GroES type chaperonins since a GroES analogue has
to date not been detected in Archaea. There is growing evidence that this class
of proteins might be involved in other functions in the cell [105, 106]. At concentrations ≥0.5 mg/ml purified *Sulfolobus* chaperonins form filaments in the

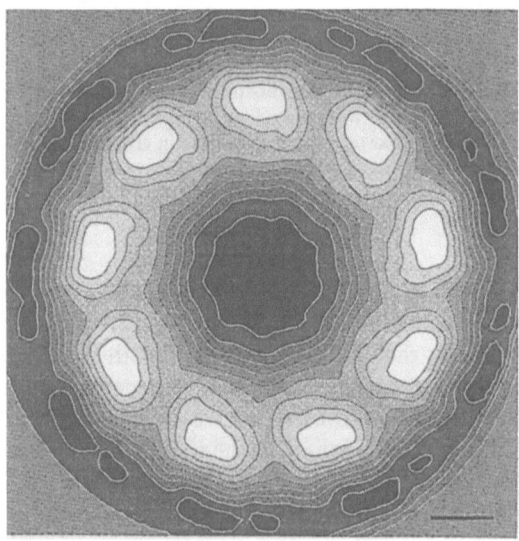

a

Fig. 15a. Two-dimensional reconstruction of the chaperonin from *Sulfolobus solfataricus*;
Average of 280 end-on views after imposing 9-fold rotational symmetry; diameter 160 Å, diameter of the central cavity 45 Å

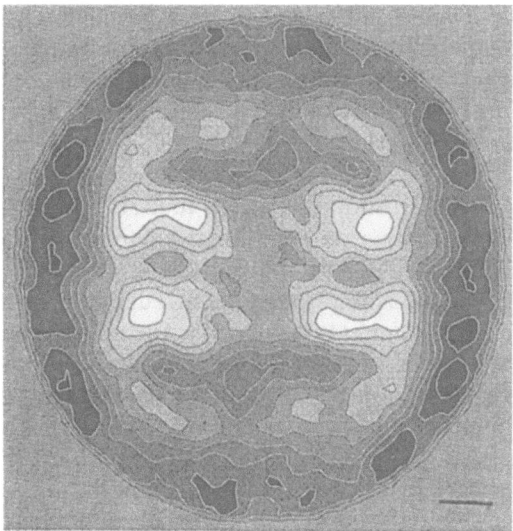

b

c

Fig. 15b–c. b Average of 23 side views with imposed 2-fold rotational symmetry; particle heigth 175 Å; (figs. 15a, 15b taken from [97]). **c** Electron microscopic projection of a negatively stained sample of a two-dimensional crystal of the *Sulfolobus solfataricus* chaperonin; 2D crystals obtained by lipid bilayer crystallization; the projection shows that the rotational symmetry of the particle is only three-fold; (fig. 15c taken from [109])

presence of Mg^{2+} and nucleotides [107]. Filament formation has been detected at physiological temperatures in biologically relevant buffers. These observations have suggested that chaperonin filaments may exist in vivo and could form extensive cytoskeletal structures. The existence of chaperonin filaments in vivo suggests furthermore a mechanism for the regulation of protein folding activities.

The side views (Fig. 15B) show the typical four-striation pattern characteristic for molecular chaperonins and suggest a two domain structure of the TF55 monomers. Reassembly studies have shown that complete TF55 molecules only reassemble if both types of subunits are present. Furthermore, this reaction requires the presence of ATP and Mg^{2+} ions [97]. The oligomers from *Sulfolobus solfataricus* are extremely resistant against dissociation even when treated with denaturing agents such as urea (8 M urea, 20 °C) at room temperature, but they dissociate into subunits at elevated temperatures. The insensitivity to acid treatment, leading to dissociation only at pH < 2 is quite remarkable. Dramatic structural changes have been observed upon phosphorylation of the *Sulfolobus solfataricus* chaperonin by ATP [97]. The phosphorylated complex has a distorted appearance in electron micrographs and the central cavity seems to be at least partly closed. The functional significance of these structural changes, however, is unclear; but a phosphorylated intermediate of the complex is probably involved in the process that leads to refolding of denatured proteins.

TF55 from *Sulfolobus solfataricus* has been crystallized from polyethylene glycol solutions in the presence of ADP and ATP. Large hexagonal prisms were obtained, which unfortunately did not diffract X-rays [108]. Well ordered 2D crystals of the *Sulfolobus solfataricus* chaperonin obtained by lipid bilayer crystallization have been described [109]. This opens up new possibilities for a 3D electron microscopic reconstruction of the particle at a resolution of around 10 Å or even better.

The crystal structure of the substrate binding domain of the α subunit of the Thermoplasma acidophilum thermosome has recently been determined by X-ray crystallography at 2.3 Å resolution [125]. Based on the established domain boundaries a soluble fragment comprising the putative substrate binding region of the α subunit has been cloned, expressed in E. coli and crystallized by vapour diffusion. A remarkable feature of the crystals was the strong polymorphism with regard to the unit cell dimensions, which is indicative of an unusual flexibility of the substrate binding domain.

The core of the protein resembles the apical domain of GroEL. A large hydrophobic surface patch has been found which is exposed by a novel helix-turn-helix motif. Structural models of the holo-chaperonin derived from cryo electronmicroscopic data point to a dual role of this helical protrusion in substrate binding and controlling access to the central cavity, obviously without the need of a GroES-like co-chaperonin.

6
The Biotechnological Significance of Enzymes from Extremophiles

Due to the high biodiversity of extremophiles and their ability to produce a large number of novel enzymes they have attracted the attention of academia and industry. Compared to mesophilic bacteria growth and productivity of enzymes, however, is often severely limited. By applying modern technologies such as molecular biology and genetic engineering it has been possible to produce enzymes from extremophiles at high concentration and analyse their structure and function in more detail. The cloning and expression of these enzymes in mesophilic hosts opens new possibilities for novel industrial processes. The following examples describe the potential use of few selected extracellular and intracellular enzymes in industrial processes. More detailed information on the enzymology and physiology of extremophiles is provided in the review articles [110–115].

6.1
Amylase and Pullulanase

Starch is found in many plants as a food reserve polysaccharide and it represents a carbon as well as an energy source for various microorganisms. This macromolecule is composed of two high-molecular weight compounds, namely amylose (15–25%) and amylopectin (75–85%), but the size and the shape of starch granules are characteristic of the plant. The importance of the starch industry is demonstrated by the large amount of sugars and sweeteners produced world wide every year. Several successive steps involving different microbial enzymes are required during the manufacture of sugars [115]. The process is initially carried out at extremely high temperature (95–105°C) and at pH 6.0–6.5. In the second step all process conditions have to be changed to a pH of 4.5 and temperatures of 60°C, due to the absence of more suitable enzymes. The discovery of more stable and specific enzymes would be significant for the improvement of the starch conversion process. Using thermostable enzymes (α-amylase, pullulanase, glucoamylase, xylose isomerase) from hyperthermophiles it will be possible to run the liquefaction and saccharrification process in one step and under the same conditions. This will also reduce the costs by avoiding the use of expensive ion exchangers. The most thermostable α-amylases and pullulanases (temperature optimum 100–105°C) have been found in the archaea *Pyrococcus woesei*, *Pyrococcus furiosus*, *Desulfurococcus mucosus*, *Pyrodictium abyssi* and *Staphylothermus marinus* [114]. The genes for the extracellular amylase and pullulanase from *Pyrococcus* sp. have been already cloned and expressed in *E. coli* and *Bacillus subtilis* [116]. *Pyrococcus woesei* pullulanase has been crystallized by vapor diffusion [124]. The crystals diffract to better than 3 Å resolution. There is an ongoing search for suitable heavy atom derivatives in order to determine the 3D structure of this enzyme. Other heat-stable amylolytic enzymes (optimum at 90°C) have also been characterised from bacteria belonging to the genera *Dictyoglomus* and *Thermotoga* [114].

6.2
Cyclodextrin Glycosyl Transferase (CGTase)

CGTase attacks α-1.4-linkages in polysaccharides by random fashion and converts starch by an intramolecular transglycosylation reaction. The nonreducing cyclization products of this reaction are α-, β, or γ-cyclodextrin consisting of 6,7, or 8 glucose molecules, respectively. The predominant application of CGTase occurs in the industrial production of cyclodextrins. Due to the ability of cyclodextrins to form inclusion complexes with a variety of organic molecules, cyclodextrins improve the solubility of hydrophobic compounds in aqueous solution. Cyclodextrin production is a multistage process in which starch is first liquefied by a heat-stable amylase and in the second step a heat-labile CGTase from *Bacillus* sp. is used. Due to the low stability of the latter enzyme the process must be run at lower temperatures. The finding of heat-stable and more specific CGTases from extremophiles will significantly improve the process. The application of heat-stable CGTases in the jet cooking, where temperatures up to 105 °C are achieved, will allow that the liquefaction and cyclization take place in one step. Thermostable CGTases are produced by the members of the genus *Thermoanaerobacter* and *Thermoanaerobacterium thermosulfurogenes* [117, 118]. Recently a heat- and alkali-stable CGTase (65 °C, pH 4–10) was purified from a newly isolated strain *Thermalkalibacter bogoriae* [119].

6.3
Xylanases

Xylan is the major component of plant hemicelluloses consisting of a main chain of β-1,4-linked D-xylopyranosyl residues. The depolymerization action of endoxylanase results in the conversion of the polymeric substrate into xylooligosaccharides. One possible use of heat-stable xylanases is the production of animal food. Thermoactive xylanases in combination with cellulases can be used for the effective conversion of these polymers to xylose and glucose. The second area of application involves the use of thermostable xylanolytic enzymes as pre-bleaching agents for kraft pulp. Here the use of xylanases will help in reducing the kappa numbers (measure of residual lignin content) of the pulp, thus reducing the requirements for chlorine during pulp bleaching. Xylanases that are optimally active at temperatures above 100 °C were detected in *Pyrodictium abyssi*, *Thermotoga thermarum*, *Thermotoga neapolitana*, *Thermotoga maritima* and *Thermotoga* sp. strain FjSS3-B.1. The genes from the latter two organisms have been cloned and expressed in *E. coli* [114].

6.4
Proteolytic enzymes

Proteinases are involved in the conversion of proteins to amino acids and peptides. The amount of proteolytic enzymes produced world wide on a commercial scale is the largest. Serine alkaline proteinases (subtilisin) are used as additives to household detergents for laundering, where they have to resist denaturation by detergents and alkaline pH. Proteinases showing high keratonolytic and elastolytic activities are used for soaking in the leather industry. Proteinases are also used as catalysts for peptide synthesis using their reverse reaction. The exploration of proteinases that can catalyse reaction under extreme conditions (high temperatures and extremes of pH) will be of value for various applications mentioned above. It has been found that most enzymes from extremophiles are stable even under the presence of high concentrations of detergents and denaturing agents. A variety of thermoactive proteinases have been identified from thermophilic archaea which belong to the genera *Pyrococcus, Thermococcus, Staphylothermus, Desulfurococcus* and *Sulfolobus.* Optimal activities have been detected between 90 and 110°C and pH 2.0 to 10 [110]. A keratin degrading thermophilic bacterium (*Fervidobacterium pennavorans*) was isolated from hot springs of the Azores Island and is able to convert chicken feather completely to amino acids and peptides at 80°C within 48 h [120].

6.5
DNA-Polymerases

DNA polymerases are the key enzymes in the replication of cellular information present in all forms of life. A variety of these enzymes, involved either in replication or repair, was characterized at the beginning of 1950 together with DNA polymerase I [121]. Thermostable DNA polymerases have played a major role in genetic engineering since the development of the polymerase chain reaction (PCR) [122]. Extremophilic microorganisms that are able to produce heat-stable DNA-polymerases with different features are attractive candidates since several problems in PCR application are still unsolved, e.g. reverse transcription at high temperatures or error-free amplification.

Various thermostable DNA-polymerases were identified in the hyperthermophiles *Pyrococcus furiosus* and *Thermococcus litoralis* and the thermophile *Thermus aquaticus.* All of these enzymes are already commercially available. Recently, a hyperthermophilic archaeon *Thermococcus aggregans* was isolated from deep-sea and was found to form a heat-stable α-DNA polymerase [123]. The gene was cloned and expressed in *E. coli.* Within the gene, three intervening sequences coding for "inteins" were identified. The presence of exonuclease activity in combination with its ability of polymerisation at high temperatures makes this DNA polymerase an attractive candidate for high PCR performance resulting in amplificates with high proofreading capacity.

7
Conclusions

Investigations of the thermostability of proteins have revealed a variety of potential stabilizing factors. Their relative importance in stabilizing proteins from hyperthermophiles is of considerable theoretical and practical interest, since these forces are responsible for stability in all proteins. It should be noted, however, that experimental structure determination, as opposed to homology modelling, is clearly needed. In one case homology modelling was used to analyze the structure of an enzyme [83] which has recently been determined by X-ray crystallography [47] but without seeing the key difference which is now obvious. Another critical point for the interpretation of differences from structural comparisons may result from the fact that hyperthermophilic enzymes, which often function only at increased temperatures, are examined at room temperature or below in crystallographic experiments or by NMR. There is certainly a need for the investigation of hyperthermostable proteins at elevated temperatures, which is an easy task for a NMR experimentalist, but more difficult to achieve in X-ray data collection. Otherwise, determinants of hyperthermostability related to structure and mobility may remain unrecognized. Hyperthermostable enzymes are often inactive at the temperatures which are used for the determination of their structures.

The mesophilic, thermophilic and hyperthermophilic enzymes which are compared in this paper demonstrate that hyperthermostability can be achieved without requiring any new types of interactions or secondary structure elements to stabilize the folded conformation. Rather than being the consequence of one dominant type of interaction it appears, that the adaptation and optimization processes, which hyperthermophilic proteins have undergone during their evolution, reflect a number of subtle interactions characteristic for each protein species that minimize the surface energy and the hydration of exposed apolar groups while burying hydrophobic residues and maximizing packing of the core and the energy due to charge-charge interactions and hydrogen bonds. It is now well documented by structure comparisons, that many thermophilic and in particular hyperthermophilic proteins show a statistically increased number of surface saltbridges and saltbridge networks. In which way these interactions contribute to thermodynamic and functional stability has still to be determined.

The sometimes tremendously increased synthesis of chaperonins in archaebacteria under heat-shock conditions suggests an involvement of these folding catalysts in the repair and refolding of proteins which have undergone damage by heat. The underlying mechanisms are still unknown. Their analysis on a molecular level is largely dependent on high-resolution structure information.

Acknowledgements. Without the dedicated work of the members of the Huddinge extremophiles group – Dr. P. Christova, Ziba Fakoor-Biniaz, Dr. Maria Flocco, Dr. Andrej Karshikoff, Dr. Stefan Knapp, Bin Ren, Dr. Velin Spassov and Gudrun Tibbelin – this review article would have never been written. The financial support of the Commission of the European Community (EC projects Biotechnology of Extremophiles, Contract BIO 2-CT93-0274 and Extremophiles as Cell Factories, Contract BIO4-CT96-0488) is gratefully acknowledged. Thanks are also due to the Deutsche Forschungsgemeinschaft and Fonds der Chemischen Industrie.

References

1. Woese CR, Kandler O, Wheelis ML (1990) Proc Natl Acad Sci USA 87:4576
2. Woese CR (1993) Biochemistry of Archaea. In: Kates M, Kushner DJ, Metheson AT (eds) New Comprehensive Biochemistry vol 26, Elsevier, Amsterdam, p 7
3. Stetter KO (1982) Nature 300:258
4. Bernhardt G, Lüdemann HD, Jaenicke R, König H, Stetter KO (1984) Naturwissenschaften 71:583
5. Jaenicke R (1991) Eur J Biochem 202:715
6. Böhm G, Jaenicke R (1994) Int J Pept Prot Res 43:97
7. Pace CN (1992) J Mol Biol 226:29
8. Knapp S, De Vos W, Rice D, Ladenstein R (1996) J Mol Biol 267:916–932
9. Macedo-Ribeiro S, Darimont B, Sterner R, Huber R (1996) Structure 4:1291
10. Pfeil W (1986) Unfolding of proteins. In: Hinz HJ (ed) Thermodynamic Data for Biochemistry and Biotechnology, Springer, Berlin Heidelberg New York, p 349
11. Pace CN, McNutt M, Gajiwala K (1996) FASEB J 10:75
12. Fersht AR (1972) J Mol Biol 64:497
13. Spassov VZ, Karshikoff AD, Ladenstein R (1994) Protein Sci 3:1556
14. Argos P, Rossmann MG, Gran UM, Zuber H, Frank G and Tratschin JD (1979) Biochemistry 18:5698
15. Stellwagen E, Wilgus H (1978) Nature 275:342
16. Spassov VZ, Atanasov BP (1994) Proteins Struct Funct Genetics 19:222
17. Spassov VZ, Karshikoff AD, Ladenstein R (1996) unpublished results
18. Ponnuswamy PG (1993) Prog Biophys Mol Biol 59:57
19. Rose GD, Wolfenden R (1993) Annu Rev Biophys Biomol Struct 22:381
20. Privalov PL, Makhadatze GI (1993) J Mol Biol 232:660
21. Hirono S, Liu Q, Moriguchi I (1991) Chem Pharm Bull (Tokyo) 39:3106
22. Tunon I, Silla E, Pascual-Ahuir JL (1992) Protein Eng 5:715
23. Spassov VZ, Karshikoff AD, Ladenstein R (1995) Protein Sci 4:1516
24. Chothia C (1974) Nature 248:338
25. Eisenberg D, McLachlan AD (1986) Nature 319:199
26. Jaenicke R (1996) FASEB J 10:84
27. Chan MK, Swarnalatha M, Kletzin A, Adams MWW, Rees DC (1995) Science 267:1463
28. Kim J, Rees D C (1992) Nature 360:553
29. Richards FM (1977) Annu Rev Biophys Bioeng 6:151
30. Miller S, Janin J, Lesk AM, Chothia C (1987) J Mol Biol 196:641
31. Richards FM, Lim WA (1994) Q Rev Biophys 26:423
32. Rashin A, Iofin M, Honig B (1986) Biochemistry 25:3619
33. Eriksson A, Base W, Zhang XJ, Heinz D, Blaber M, Baldwin E, Matthews B (1993) Science 255:178
34. Ishikawa K, Nakamura H, Morikawa K, Kanaya S (1993) Biochemistry 32:6171
35. Sutherland KJ, Henneke CM, Towner P, Hough DW, Danson MJ (1990) Eur J Biochem 194:839
36. Sutherland KJ, Danson MJ, Hough DW, Towner P (1991) FEBS Lett 282:132
37. Russel RJM (1994) The crystal structure of *Thermoplasma acidophilum* citrate synthase, Ph. D. Thesis, University of Bath, UK
38. Remington SJ, Wiegand G, Huber R (1982) J Mol Biol 158:111
39. Russel RJM, Hough D, Danson MJ, Taylor GL (1994) Structure 2:115
40. Kleywegt GJ, Jones TA (1994) Acta Crystallogr. D50:178
41. Karshikoff A, Ladenstein R (1997) unpublished results
42. Day MW, Hsu BT, Joshua-Tor L, Park JB, Zhou ZH, Adams MWW, Rees DC (1992) Protein Sci 1:1494
43. Blake PR, Park JB, Zhou ZH, Hare DR, Adams MWW, Summers MF (1992) Protein Sci 1:1508

44. Busse SA, La Mar GN, Yu LP, Howard JB, Smith ET, Zhou ZH, Adams MWW (1992) Biochemistry 31:11952
45. Macedo-Ribeiro S, Darimont B, Sterner R, Huber R (1996) Structure 4:1291
46. Korndoerfer I, Steipe B, Huber R, Tomschy A, Jaenicke R (1995) J Mol Biol 246:X511
47. Yip K, Stillman TJ, Britton KL, Artymiuk PJ, Baker PJ, Sedelnikova SE, Engel PC, Pasquo A, Chiaraluce R, Consalvi V, Scandurra R, Rice DW (1995) Structure 3:1147
48. Tomschy A, Glockshuber R, Jaenicke R (1993) Eur J Biochem 214:43
49. Stillman TJ, Baker PJ, Britton KL, Rice DW (1993) J Mol Biol 234:1131
50. Eggen RIL, Geerling ACM, Waldkötter K, Antranikian G, de Vos WM (1993) Gene 132:143
51. Rossmann MG, Moras D, Olsen KW (1974) Nature 250:194
52. Hennig M, Darimont B, Sterner R, Kirschner K, Jansonius JN (1995) Structure 3:1295
53. Consalvi V (1996) personal communication.
54. Knapp S (1996) personal communication
55. Ishikawa K, Okumura M, Katayanagi K, Kimura S, Kanaya S, Nakamura H, Morikawa K (1993) J Mol Biol 230:529
56. Kelly CA, Nishiyama M, Ohnishi Y, Beppu T, Birktoft JJ (1993) Biochemistry 32:3913
57. Perutz MF, Raidt H (1975) Nature 255:256
58. Horovitz A, Serrano L, Avron B, Bycroft M, Fersht AR (1990) J Mol Biol 216:1031
59. Waldburger CD, Schildbach JF, Sauer RT (1995) Nature Struct Biol 2:122
60. Carter PJ, Winter G, Wilkinson AJ, Fersht AR (1984) Cell 38:835
61. Horovitz A, Fersht AR (1990) J Mol Biol 214:613
62. Privalov P (1982) Adv Prot Chem 35:1
63. Pace CN (1986) Methods Enzymol 131:266
64. Baumann H, Knapp S, Lundbäck T, Ladenstein R, Härd T (1994) Nature Struct Biol 11:808
65. Program DALI, Holm L, Sander C (1993) J Mol Biol 233:123
66. Edmondson SP, Qiu L, Shriver JW (1995) Biochemistry 34:13289
67. Dijk J, Reinhardt R (1986) The structure of DNA binding proteins from eu and archae-bacteria. In: Gualerzi CO, Pon CL (eds) Bacterial Chromatin. Springer, Berlin Heidelberg New York, p 185
68. Knapp S, Karshikoff A, Berndt KD, Christova P, Atanasov B, Ladenstein R (1996) J Mol Biol 264:1132–1144
69. Klump H, DiRuggiero J, Kessel M, Park J, Adams MWW, Robb FT (1992) J Biol Chem 267:22681
70. McAfee JG, Edmondson SP, Datta PK, Shriver JW, Gupta R (1995) Biochemistry 34:10063
71. Becktel, Schellman. 1987, Biopolymers 26, 1859
72. Swint L, Robertson AD (1993) Prot Sci 2:2037
73. Alexander P, Fahnestock S, Lee T, Orban J, Bryan P (1992) Biochemistry 31:3597
74. Privalov PL, Makhatadze GI (1990) J Mol Biol 213:385
75. Gomez J, Hilser VJ, Xie D, Freire E (1995) Proteins 22:404
76. Nojima H, Ikai A, Oshima T, Noda H (1977) J Mol Biol 116:429
77. Day MW, Hsu BT, Joshua-Tor L, Park JB, Zhou ZH, Adams MWW, Rees DC (1992) Prot Sci 1:1494
78. Privalov PL, Gill SJ (1988) Adv Prot Chem 39:191
79. Murphy KP, Privalov PL, Gill SJ (1990) Science 247:559
80. Ragone R, Colonna G (1995) J Am Chem Soc 117:16
81. Griko YV, Makhatadze GI, Privalov PL, Hartley RW (1994) Protein Sci 3:669
82. Yu Y, Makhatadze GI, Pace N, Privalov PL (1994) Biochemistry 33:3312
83. Britton KL et al.,, Yip KSP (1995) Eur J Biochem 229:688
84. Risse B, Stempfer G, Rudolph R, Jaenicke R (1992) Protein Sci 1:1699
85. Timasheff SN (1993) Annu Rev Biophys Biomol Struc 22:67
86. Santoro MM, Liu Y, Khan SMA, Hou LX, Bolen DW (1992) Biochemistry 31:5278
87. Ellis RJ, van der Vies SM (1991) Annu Rev Biochem 60:321
88. Hendrick JP, Hartl FU (1995) FASEB J 9:1559

89. Trent JD, Osipiuk J, Pinkau T (1990) J Bacteriol 172:1478
90. Ellis RJ (1992) Nature 358:191
91. Phipps BM, Hofmann A, Stetter KO, Baumeister W (1991) EMBO J 10:1711
92. Phipps BM, Typke D, Hegerl R, Volker S, Hoffmann A, Stetter KO, Baumeister W (1993) Nature 361:475
93. Marco S, Urena D, Carrascosa JL, Waldmann T, Peters J, Hegerl R, Pfeifer G, Sack-Kongehl H, Baumeister W (1994) FEBS Lett 341:152
94. Waldmann T, Nimmesgern E, Nitsch M, Peters J, Pfeifer G, Muller S, Kellermann E, Engel A, Hartl FU, Baumeister W (1995) Eur J Biochem 227:848
95. Lewis VA, Hynes GM, Zheng D, Saibil H, Willison K (1992) Nature 358:249
96. Frydman J, Nimmesgern E, Erdjument-Bromage H, Wall S, Tempst P, Hartl FU (1992) EMBO J 11:4767
97. Knapp S, Schmidt-Krey I, Hebert H, Bergman T, Jörnvall H, Ladenstein R (1994) J Mol Biol 242:397
98. Trent JD, Nimmesgern E, Wall JS, Hartl FU, Horwich AL (1991) Nature 354:490
99. Martin J, Langer T, Boteva R, Schramel A, Horwich AL, Hartl FU (1991) Nature 352:36
100. Carrascosa JL, Abella G, Marco S, Carazo JM (1990) J Struct Biol 104:2
101. Braig K, Otwinowski Z, Hedge R, Boisvert DJ, Joachimiak A, Horwich AL, Sigler PB (1994) Nature 371:578
102. Langer T, Pfeifer G, Martin J, Baumeister W, Hartl FU (1992) EMBO J 11:4757
103. Saibil H, Wood S (1993) Curr Op Struct Biol 3:207
104. Guagliardi A, Cerchia L, Bartolucci S, Rossi M (1994) Protein Sci 3:1436
105. Ursic D, Sedbrook JC, Himmel KL, Culbertson MR (1994) Mol Cell Biol 5:1065
106. Brown CR, Doxsey SJ, Hong-Brown LQ, Martin RL, Welch WJ (1996) J Biol Chem 271:824
107. Trent JD, Kagawa HK, Yaoi T, Olle E, Zaluzec NJ (1997) Proc Natl Acad Sci USA 94:5383–5388
108. Knapp S, Ladenstein R (1995) unpublished results
109. Ellis M, Knapp S, Koeck PJB, Fakoor-Biniaz Z, Ladenstein R, Hebert H (1997) manuscript submitted
110. Leuschner C, Antranikian G (1995) World. J Microbiol Biotechnol 11:95
111. Krahe M, Antranikian G, Märkl H (1996) FEMS Microbiol Reviews 18:271
112. Stetter KO (1996) FEMS Microbiol Reviews 18:149
113. Sunna A, Antranikian G (1997) Crit Rev Biotechnology 17:39
114. Sunna A, Moracci M, Rossi M, Antranikian G (1997) Extremophiles 1:2
115. Antranikian G (1992) Microbial degradation of starch. In: Winkelmann G (ed) Microbial degradation of natural products. VCH, Weinheim, p 27
116. Jorgensen S, Vorgias C, Antranikian G (1997) J Biol Chem in press
117. Petersen S, Jensen B, Dijkhuizen L, Jorgensen S, Dijkstra B (1995) Chemtech 12:19
118. Wind R, Liebl W, Buitelaar R, Penninga D, Spreinat A, Dijkhuizen L, Bahl H (1995) Appl Environ Microbiol 61:1257
119. Prowe S, Van de Vossenberg J, Driessen A, Antranikian G, Konings W (1996) J Bacteriol 178:4099
120. Friedrich A, Antranikian G (1996) Appl Environ Microbiol 62:2875
121. Kornberg A, Baker T (1992) DNA Replication. 2nd edn WM Freeman, New York
122. Mullis K, Faloona F, Saiki R, Horn G, Ehrlich H (1986) Specific enzymatic amplification of DNA in vitro: the polymerase chain reaction. Cold Spring Harbour Sym. Quant Biol 247:7116
123. Canganella F, Jones WJ, Gambacorta A, Antranikian G (1997) Arch Microbiol 167:233
124. Knapp S, Rüdiger A, Antranikian G, Jorgensen PL, Ladenstein R (1995) Proteins 23:595
125. Klumpp M, Baumeister W, Essen LO (1997) Cell 91:263

Received August 1997

Molecular Biology of Hyperthermophilic *Archaea*

John van der Oost[1]* · Maria Ciaramella[2] · Marco Moracci[2] · Francesca M. Pisani[2] · Mose Rossi[2] · Willem M. de Vos[1]

[1] Department of Microbiology, Wageningen Agricultural University,
  Hesselink van Suchtelenweg 4, 6307 CT Wageningen, The Netherlands
  E-mail: john.vanderoost@algemeen.micr.wau.nl
[2] Institute of Protein Biochemistry and Enzymology, Via Marconi 10, 80125 Naples, Italy

The sequences of a number of archaeal genomes have recently been completed, and many more are expected shortly. Consequently, the research of *Archaea* in general and hyperthermophiles in particular has entered a new phase, with many exciting discoveries to be expected. The wealth of sequence information has already led, and will continue to lead to the identification of many enzymes with unique properties, some of which have potential for industrial applications. Subsequent functional genomics will help reveal fundamental matters such as details concerning the genetic, biochemical and physiological adaptation of extremophiles, and hence give insight into their genomic evolution, polypeptide structure-function relations, and metabolic regulation. In order to optimally exploit many unique features that are now emerging, the development of genetic systems for hyperthermophilic *Archaea* is an absolute requirement. Such systems would allow the application of this class of *Archaea* as so-called "cell factories" : (i) expression of certain archaeal enzymes for which no suitable conventional (mesophilic bacterial or eukaryal) systems are available, (ii) selection for thermostable variants of potentially interesting enzymes from mesophilic origin, and (iii) the development of *in vivo* production systems by metabolic engineering. An overview is given of recent insight in the molecular biology of hyperthermophilic *Archaea*, as well as of a number of promising developments that should result in the generation of suitable genetic systems in the near future.

Keywords: Hyperthermophile, Archaea, molecular biology, transcription, translation, regulation

* Corresponding author.

Advances in Biochemical Engineering /
Biotechnology, Vol. 61
Managing Editor: Th. Scheper
© Springer-Verlag Berlin Heidelberg 1998

1
Introduction

Two decades ago it was realized that the division of life on earth into the kingdoms of the Prokaryotes and the Eukaryotes, as proposed by Stanier and Van Niel [1], should be revised. Carl Woese and coworkers were pioneering with the application of molecular biological techniques for classification purposes. They concluded that the Prokaryotes are actually composed of two distinct phylogenetic entities, the *Bacteria* (eubacteria) and the *Archaea* (archaebacteria) that, together with the *Eukarya,* comprise the three "domains of life" [2, 3]. Whereas Woese's classification was originally based on 16S rRNA analysis, it has since been confirmed by phylogenetic studies of a variety of ubiquitous macromolecules, a.o. 23S rRNA, DNA-dependent RNA polymerase, translation factors, ribosomal proteins, ATPases (reviewed in [4]). *Archaea* are now commonly regarded as a monophyletic group, i.e. originating from a single common ancestor. An early split gave rise to two distinct archaeal orders, the *Euryarchaeota* (methanogens, halophiles, (hyper)thermophiles) and the *Crenarchaeota* (sulphate-reducing (hyper)thermophiles) (Table 1).

The aim of the present paper is to integrate many features that in one way or another contribute to the process of proteins synthesis in hyperthermophilic *Archaea*. As defined by Karl Stetter *et al.* [5], hyperthermophiles have an optimum growth temperature above 80 °C, which apparently includes many *Archaea* and in addition a number of *Bacteria* (Table 1). When these hyperthermophiles were first isolated in the early 1980s, the question arose as to how these organisms were able to survive under such harsh conditions. A major issue being dealt with in the research of hyperthermophiles concerns the enhanced

Table 1. Taxonomy and optimum growth temperature of (hyper)thermophilic *Archaea* and *Bacteria* that are discussed in the text [5, 10]

Order	Species	Opt. temp. (°C)
Archaea – Euryarchaeota		
Thermococcales	*Thermococcus litoralis*	88
	Pyrococcus furiosus	100
	Pyrococcus abyssi	97
Methanopyrales	*Methanopyrus kandleri*	98
Methanococcales	*Methanococcus jannaschii*	80
	Methanobacterium thermoautotrophicum	65 – 70
	Methanococcus thermolithotrophicus	65 – 70
	Methanococcus vannielii	35 – 40
Archaea – Crenarchaeota		
Sulfolobales	*Sulfolobus acidocaldarius*	75
	Sulfolobus solfataricus	80
	Sulfolobus shibatae	81
	Desulfurolobus ambivalens	81
Desulfurococcales	*Desulfurococcus mobilis*	85 – 88
Pyrodictiales	*Pyrodictium occultum*	105
Thermoproteales	*Thermoproteus tenax*	88
	Pyrobaculum aerophilum	100
Bacteria		
Thermotogales	*Thermotoga maritima*	80
Aquificales	*Aquifex pyrophilus*	85

stability of biological macromolecules (proteins, nucleic acids and lipids), addressing fundamental questions such as structure-function relations at extreme conditions (reviewed in [6 – 9]). In addition, hyperthermophile research has been inspired significantly by the potential applications of thermostable enzymes in a range of industrial processes.

Here we focus on some novel developments that relate to the different levels of protein synthesis in hyperthermophilic archaea: chromosome topology, gene organization, transcription, translation, and regulatory features. In order to establish certain phylogenetic relations, we compare the different archaeal processes and components to well-established analogous systems in *Bacteria* and *Eukarya*. When experimental data are available, the comparison will be made between mesophilic and (hyper)thermophilic classes of *Archaea*, to reveal features that might contribute to enhanced thermostability.

2
Chromosome Structure and Replication

Archaeal genomes are organized as circular chromosomes, occasionally supplemented with extrachromosomal DNA molecules. The topological state of archaeal DNA molecules appears to be mediated by a variety of polypeptides:

topoisomerases (positive/negative coiling state), histone-like proteins (packaging, supercoiling), DNA polymerase (unwinding and replication), and RNA polymerase (local unwinding and transcription). All these archaeal features resemble the bacterial genome organization.

2.1
Thermostability

The stabilization of DNA at elevated temperatures, i.e. preventing the helices to separate, theoretically may be brought about by intrinsic or extrinsic factors [9]. Apparently, there is no clear correlation between the G+C content of a genome and the optimum growth temperature. Rather, stabilization of chromosomal DNA appears to be brought about by the presence of extrinsic factors, e.g. polyamines, potassium ions, histone-like proteins.

In contrast to the genomic DNA, the sequences of 16 S and 23 S rRNA genes of thermophiles does show a significantly elevated G+C content. In addition, it has been noted that, in the large rRNA genes of hyperthermophiles, the relative increase in cytosine content is greater than that of guanine, which might suggest that G-U pairs are substituted by G-C pairs [10]. Although the G+C content of open reading frames is proportional to that of the genome, there appears to be a tendency that the C content increases more rapidly than the G content [10], possibly contributing to an increased transcript stability.

Histone-like proteins may be involved in enhancing the thermostability of DNA as well. It has been demonstrated that DNA in nucleosome-like structures is positively supercoiled [11]. Based on this observation it was proposed that, apart from contributing to the DNA thermostability, a physiological role of the histone homologues might be to facilitate transcription by relaxation of non-covered DNA regions.

2.2
Topoisomerases

In bacterial nucleoprotein as well as in eukaryal chromatin, DNA apparently has a negative superhelicity. The analysis of the topological state of chromosomal DNA in hyperthermophilic *Archaea* is just emerging, and many questions still remain unanswered [12]. An interesting observation is the fact that, at least in some cases, DNA molecules in hyperthermophiles might be positively supercoiled. This is based on the observation that DNA of the archaeophage SSV1, and of several plasmids from hyperthermophilic species is supercoiled in vivo [13]. Moreover, a peculiar topoisomerase activity has been found in hyperthermophiles [14], which has been called "reverse gyrase", since it catalyzes *in vitro* positive supercoiling of a closed circular DNA at high temperature [15]. This enzymatic activity is widely distributed among (hyper)thermophilic *Archaea* [16, 17], as well as thermophilic bacterial species [18]. The best-characterized reverse gyrase is the one purified from the thermoacidophilic archaeon *Sulfolobus acidocaldarius* [19]. This enzyme, a monomer with a molecular mass of 130 kDa, is a type I topoisomerase since it introduces a transient single-stranded DNA break and forms a reaction intermediate by covalently binding the 5'-end of the break, as reported for bacterial DNA topoisomerase I [20, 21]. However, unlike the alternative bacterial and eukaryal type I DNA topoisomerases, its activity depends on ATP. Recently the gene coding for *S. acidocaldarius* reverse gyrase has been cloned and sequenced [22]. Analysis of the primary structure revealed that its N-terminal half contains several helicase motifs, including an ATP-binding site, whereas the C-terminal portion is related to bacterial topoisomerase I and to *Saccharomyces cerevisiae* Top 3 [23]. Based on these results, a model has been proposed in which positive gyration is the net result of a helicase-plus-topoisomerase I process. The biological function of reverse gyrase is not yet clearly understood, although positive supercoiling occurs *in vivo* in thermophilic *Archaea* [24]. However, the finding that this activity is present in all thermophilic species of *Archaea* and *Bacteria* tested suggests that it could play a role in the stabilization of DNA molecules at high temperature. In addition to reverse gyrase, several other DNA topoisomerase activities have been identified in hyperthermophiles (reviewed in [9]). An ATP-independent type I topoisomerase which relaxes only negatively supercoiled DNA (Topo III) has been purified from *Desulfurococcus mobilis* [15, 25]. Type II topoisomerases able to relax both negatively or positively supercoiled DNA (Topo II) have been purified from both *Sulfolobus* and *Pyrococcus* [9, 26]. More interestingly, a 3'-topo I enzyme, called Topo V, has been purified from *Methanopyrus kandleri*. Enzymes of this type, which bind to the 3' end of a DNA break and relax both positive and negative superturns, have previously been found only in *Eukarya* [25].

2.3
Replication

Details on the mechanisms of DNA replication and repair in *Archaea* are largely unknown. Origins of replication, origin-binding factors, DNA helices, single stranded-DNA binding proteins, and DNA polymerase accessory factors have not yet been functionally identified in hyperthermophilic *Archaea*, and no *in vitro* system to study DNA replication is available.

DNA polymerases from several (hyper)thermophilic bacterial and archaeal species have been extensively characterized with respect to their exploitation in *in vitro* DNA synthesis at high temperatures, e.g. PCR and sequencing reactions. These enzymes are monomeric, with a molecular mass of 80–130 kDa, and generally are extremely thermostable, consistent with the optimal growth temperature of the microbes from which they have been isolated. A number of these DNA polymerases possess an associated 3'–5' exonuclease activity [27–31] and for some proofreading activity has been demonstrated [29, 32]. The DNA polymerase purified from *S. solfataricus* was demonstrated to possess a modular organization of its associated catalytic activities, consisting of two domains: the N-terminal domain is responsible for the 3'–5' exonuclease, whereas the C-terminal domain catalyzes the actual DNA polymerization [31, 33]. Like the eukaryal α-type DNA polymerase, the DNA polymerase purified from *S. solfataricus* appears to be sensitive to aphidicolin, suggesting some structural conservation [34]. On the other hand, no effect by this drug was observed on DNA polymerase activities isolated from *S. acidocaldarius* [27] and *M. thermoautotrophicum* [35].

The genes coding for several thermophilic archaeal DNA polymerases have been cloned and sequenced [36-39]. The analysis of protein primary structure indicates that they are all related to the B family of DNA polymerases, which includes eukaryal α and δ DNA polymerases, replicases of eukaryal viruses, as well as DNA polymerase II from *E. coli* and from phage F29 [40]. The DNA polymerase gene from the hyperthermophilic archaeon *Thermococcus litoralis* contains two in-frame intervening protein sequences (inteins) [37], which are removed post-translationally by a novel protein splicing mechanism [41]. The physiological role of these DNA polymerases in DNA replication or repair needs to be established. It is anticipated that multiple DNA polymerase activities with different biological functions exist in *Archaea,* as generally reported for *Bacteria* and *Eukarya.* Indeed, two distinct DNA polymerases of family B have been identified in *S. solfataricus* and *Pyrodictium occultum* [38, 42]. It should be noticed that the enzymes for which a detailed kinetic characterization is available show low processivity values, at least *in vitro,* suggesting that they could be involved in repair rather than *de novo* synthesis processes. However, assuming that additional factors are required *in vivo* to achieve optimal processivity, one can at this stage not rule out the participation of these DNA polymerases in replication.

3
Mobile Elements

In recent years mobile elements, including viruses, plasmids and IS elements, have been discovered in various hyperthermophilic *Archaea.* These mobile elements have great potential in developing tools for genetic expression systems as will be discussed below. Several of these genetic elements have been completely sequenced which has helped in analyzing their coding capacity and in studying signals involved in transcription and, if applicable, replication, transposition or lysogeny.

No viruses have yet been described for the hyperthermophiles belonging to the *Euryarchaeota.* However, the often observed spontaneous lysis of *Pyrococcus furiosus* at the end of the growth phase might indicate the presence of a cryptic virus [43]. In contrast, many viruses have been isolated from *Crenarcheaota,* notably from *Sulofolobus* and *Thermoproteus* spp., which have been inventoried recently by Zillig *et al.* [44]. SSV1 is the best investigated virus: it has been completely sequenced (15.4 kb), and it exists as a lysogen inserted in the arginyl tRNA gene from its host *S. shibatei* [45]. The lysogenic state of this virus seems to be controlled by a repressor and induction is evoked by UV radiation or treatment with mitomycin C, suggesting similarities with the SOS response of bacterial viruses.

Plasmids have been found in all phylogenetic groups of hyperthermophilic *Archaea.* The first plasmid described was the high copy number 7.7-kb pDL10 of *Desulfurolobus ambivalens* that is widely distributed in this member of the

Sulfolobales [46]. There are indications that pDL10 is involved in autotrophic growth of *D. ambivalens* but its coding capacity is presently unknown. In an extensive screening program a panoply of multicopy plasmids was discovered in the Sulfolobales [47]. Some of these are cryptic but others may encode distinct phenotypes such as the production of antimicrobial compounds [44]. One plasmid, the small 5.5-kb pRN1 has been sequenced and appears to code only for its own replication [47]. Of great potential is the finding that the 45-kb plasmid pNOB8 can be transferred by a process that resembles bacterial conjugation and requires cell-cell contact [48].

In the hyperthermophilic *Euryarchaeota* plasmids appear to be less abundant. A single strain of *Pyrococcus*, the deep sea isolate *Pyrococcus abyssi*, was found to contain a multicopy plasmid pGT5 (3.4 kb) that has recently been completely sequenced [49]. Remarkably, it accumulates single-stranded replication intermediates, suggesting that it replicates via a rolling circle mode of replication. The template (plus) strand of pGT5 encodes a potential replication protein with homology to bacterial Rep proteins and another protein that may be involved in recombination [49].

So far only a single IS element has been described in hyperthermophilic *Archaea*. This element of 1150 bp has structural resemblance with bacterial IS elements and has been discovered because of its transposition in the *lacS* gene in *S. solfataricus*, resulting in loss of β-galactosidase activity [50]. Recently, another class of mobile elements was discovered in hyperthermophiles, the mobile introns. It has been known for some time that ribosomal RNA's of several hyperthermophiles contain introns (see Sect.6 – Gene Organization and Regulation). The 23 S rRNA of *Desulfurococcus mobilis* contains such an intron that has been found to spread in *Sulfolobus acidocaldarius* since intron-carrying cells appear to have a selective advantage over intron-less cells [51].

The exchange of chromosomal genes from hyperthermophilic *Archaea* has recently been demonstrated. Upon mixing and plating of two distinct auxotrophic mutants of *Sulfolobus acidocaldarius*, stable genetic recombinants were obtained [51 a].

4
Transcription

All cellular RNA is synthesized by the DNA-dependent RNA polymerase according to the information stored on the genome. In addition, gene transcription generally requires the concerted action of one or more assisting polypeptides that mediate the interaction of the RNA polymerase complex with a target promoter.

4.1
Bacterial and Eukaryal Transcription

The bacterial $\alpha_2\beta\beta'$-type RNA polymerase associates with a sigma (σ) factor, resulting in the formation of a RNA polymerase-holoenzyme. Whereas *Bacteria* express only a single RNA polymerase core, they have the capacity to express a set of specific sigma factors. Upon formation of a holoenzyme complex, each sigma factor directs the RNA polymerase core to a specific promoter, the transcription start signals on a given DNA template. This implies that sigma factors play an important role in controling both the rate and the specificity of transcription. In order to accomplish proper gene regulation, it is obvious that the expression of at least some of the sigma factors themselves will be tightly controlled as well, e. g. the σ^{32} heat shock sigma factor of *E. coli*. In *Bacillus subtilis*, up to 18 distinct sigma factors have been identified, many of which are involved in the transcription initiation of a specific subset of genes and operons at different stages of sporulation [52+52a]. Two hexanucleotide sequences in the bacterial promoter region are involved in the interaction with the sigma factor linked to RNA polymerase complex, centered approximately at the −35 and the −10 position relative to the transcription start (+1) [52, 53]. The actual catalytic center, where ribonucleotide triphosphate substrate is bound and processed, is located in the β subunit, whereas the β' subunit is involved in binding the DNA during elongation. The role of the α subunit couple appears to be a regulatory one, since transcription regulators like cAMP-receptor protein (Crp) have been suggested to interact with both the α and the σ subunits of the RNA polymerase during initiation [54, 55]. Upon transcription elongation, the sigma factor dissociates again from the RNA polymerase core that moves along the coding strand. Although it is a matter of definition to classify the bacterial sigma polypeptide as an RNA polymerase subunit or rather as a transcription initiation factor, its function certainly resembles a number of features of "classical" transcription factors (see below): associating to RNA polymerase, mediating the binding of the holoenzyme to a promoter, potentially interacting with transcription regulators, and dissociating from the RNA polymerase core when the transcription of the DNA template begins. Despite these analogous features, however, the sigma factors appear not to be structurally related to any of the presently known eukaryal and archaeal transcription factors (see below) [52, 53].

Due to the lack of chromatin structure, the bacterial promoters are believed to be freely accessible to the RNA polymerase holoenzyme. In contrast, the compact nucleosome structure of the eukaryal chromosome will be far less accessible to the RNA polymerase complex. Hence, a primary event in transcription initiation will be the local dissociation of the chromatin structure. In the latter process, transcription factors (TFs) might play a role. Subsequently, the TATA-binding protein (TBP) interacts in a specific manner with the TATA box, a eukaryal promoter element centered 25–30 base pairs upstream the transcription start. The TBP is one component of a multisubunit complex TFIID, which forms a stable protein-DNA complex. After formation of the TFIID(TBP)-TATA-element complex, the subsequent event is the association of TFIIB, another general transcription factor, resulting in the TFIIB-TFIID(TBP)-DNA ternary complex (the preinitiation complex). When these polypeptides occupy the promoter, the stage is set for entry of the RNA polymerase. Together with at least one transcription initiation factor, TFIIF, the RNA polymerase completes the assembly of the initiation complex. Eukaryal *in vitro* transcription systems have been developed successfully, using a DNA fragment and three polypeptide components: TBP, TFIIB and RNA polymerase II [56]. Different experimental conditions, however, require the presence of additional transcription factors, like TFIIE, TFIIF and TFIIH. The actual transcription begins after local un- winding at the transcription start site. Upon transcription initiation, the RNA polymerase II migrates along the coding strand, thus dissociating from the tran- scription factors of the preinitiation complex (TFIID, TFIIB) that may remain bound to the promoter, allowing rapid reinitiation (reviewed in [57]). An alter- native assembly mechanism has recently been proposed in which the RNA polymerase II holoenzyme contains many, if not all, of the general initiation factors, except for TFIID. Yeast RNA polymerase II holoenzymes have been described that consist of RNA polymerase II, a subset of general transcription factors, and nine SRB-type regulatory proteins (extragenic suppressors of *S. cerevisiae* RNA polymerase II mutations). Recent evidence indicates that many of these components assemble into a large complex called the RNA polymerase holoenzyme, the SRB components of which participate in the response to trans- criptional regulators [58]. Recently, a human homologue of the yeast SRB7 gene has been isolated, and antibodies against human SRB7 protein were used to purify and characterize a mammalian RNA polymerase II holoenzyme, con- taining the general transcription factors TFIIE and TFIIH. This holoenzyme is more responsive to transcriptional activators than core RNA polymerase II when assayed in the presence of coactivators [59]. For an extensive review on RNA polymerase II transcription control, see [59a].

4.2
Archaeal Transcription

Even before the recognition of the *Archaea* being a monophyletic entity [2], it was noticed that certain (methanogenic) prokaryotes possess a number of eukaryal features. The latter observation suggested a certain relatedness between eukaryotes and these microorganisms, presently classified as *Archaea* [60, 61]. Subsequently, Zillig and coworkers corroborated this relatedness by the demonstration that certain subunits of the archaeal DNA-dependent RNA poly- merase shared significantly more homology with subunits of the eukaryal enzy- me complex, than with the relatively simple bacterial enzyme [62, 63]. Recently, an extended biochemical analysis of the *Sulfolobus acidocaldarius* RNA poly- merase resulted in establishing the primary sequences of 13 subunits. Sequence comparison and a subsequent phylogenetic analysis confirmed the closer homology of the archaeal with the three eukaryal RNA polymerase types (*Saccharomyces cerivisiae*) (Table 2) [64].

The conclusion that the *Archaea* are more akin to *Eukarya* than *Bacteria*, as deduced from the RNA polymerase phylogenetic analysis, is in line with the con- clusions drawn from the rooted phylogenetic trees, based on sequence informa- tion of elongation factors EF Tu/1α and G/2 [65], ATPase α and β subunits [66], and recently of some aminoacyl-tRNA synthetases [67] (Fig. 1A). However, a phylogenetic dendrogram of the truncated A'/ A'' RNA polymerase subunit of *Sulfolobus*, the eukaryal counterparts A190 (pol I), B220 (pol II) and C160 (pol III), and the related bacterial β', suggests that archaeal RNA polymerase is most

closely related to eukayal pol II and pol III, whereas the bacterial RNA polymerase appears to group with the eukaryal pol I [4]. It is concluded that the most
likely evolutionary scenario explaining the observed clustering would be some
sort of fusion event of unknown nature between early archaeal and possibly
bacterial genomes: pol II and pol III being derived from an archaeon, pol I from
a bacterium (or a very early archaeal ancestor). This so-called "fusion hypothesis" may offer an explanation for the fact that, apart from RNA polymerase I
[4], a substantial number of eukaryal polypeptides (e.g. Hsp70, glutamate
dehydrogenase, ferredoxin) share a significant higher degree of homology with
bacterial counterparts than with archaeal ones [68] (Fig. 1B). Zillig *et al.* [4]
concluded that "the eukaryotic cell may be more chimeric than Lynn Margulis
initially imagined" when she posed the endosymbiont theory [69].

Table 2. Comparison of the subunit composition of the archaeal DNA-dependent RNA polymerase (*S.a., Sulfolobus acidocaldarius*), with the bacterial complex (*E.c., Escherichia coli*), and
three eukaryal types (*S.c., Saccharomyces cerevisiae*), (modified after [64]). For the core subunits, the percentage of identity is indicated (%). The subunits EHKLN have counterparts in
(some) eukaryal types, some subunits from the archaeal complex have not been demonstrated
to have bacterial or eukaryal counterparts; on the other hand, some subunits/factors of the
eukaryal complexes (numbers based on data reported by [4, 64]) and the bacterial
complex (σ, sigma factor) do not have any known counterparts in *Archaea*, however see
Sect. 6 – Gene Organization and Regulation [119])

S.a. (*Archaea*)	A′–A″ (%)	B	(%)	D	(%)	EHKLN	FGI(M)	–
E.c (*Bacteria*)	β′ (30–25)	β	(25)	α	(22)	--	--	σ
S.c. pol I (*Eukarya*)	A190 (36–32)	A135	(30)	AC40	(34)	HKLN	--	6
S.c. pol II (*Eukarya*)	B220 (43–33)	B150	(44)	B44	(25)	EHKLN	--	2
S.c. pol III (*Eukarya*)	C160 (43–34)	C128	(38)	AC40	(34)	EHKLN	--	6

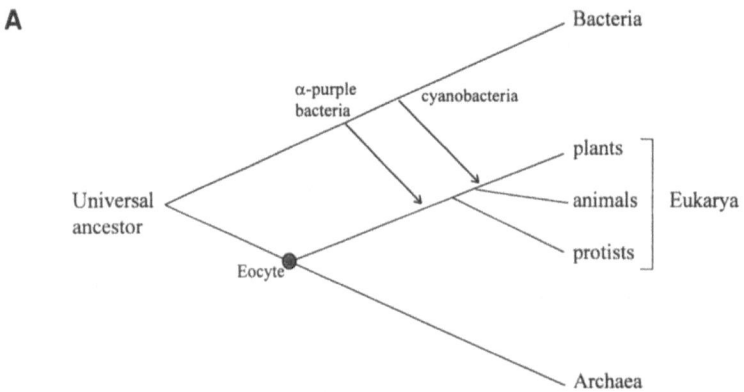

Fig. 1 A, B. Tentative scenarios on the evolutionary origin of eukaryotic cells (after Gupta and
Golding [68]): A traditional "archaebacterial model" [3, 65–67]

B

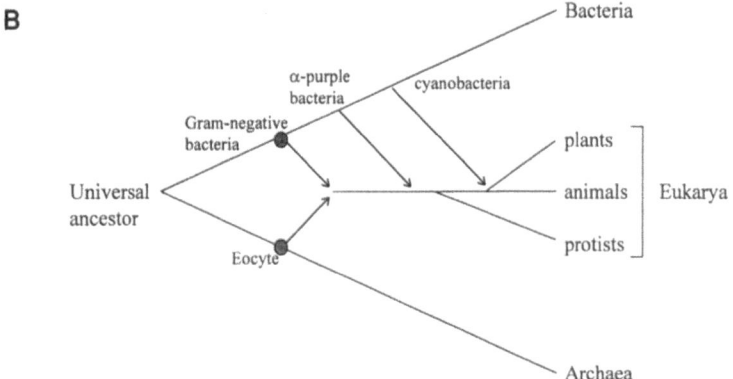

Fig. 1. B alternative chimeric (fusion) model [4, 68]. In the archaebacterial model, the eukaryotic nuclear genome is directly evolved from an archaeal ancestor, the Eocyte. In the chimeric model, however, a unique fusion event between a Gram-negative bacterium and an Eocyte archaeon gave rise to the ancestral eukaryotic cell in which both parents contributed to its nuclear genome

4.3
Archaeal Promoters and Transcription Factors

Alignment of the flanking regions of a number of genes from different *Archaea* (*Crenarchaeota* and *Euryarchaeota*) revealed the presence of a TATA motif located around a position between –30 and –25, relative to the transcription start (+1), resembling the eukaryal RNA polymerase II promoter organization [70–73]. In vitro transcription analysis of a series of mutants of a *Sulfolobus shibatae* rRNA promoter pinpointed a core promoter region between positions –2 and –38, containing all information for transcription. More detailed characterization of this region indicated two sequence elements important for promoter function: a distal one located between positions –38 and –25 (TATA box, box A) and a proximal one located between positions –11 and –2 (box B) [72]. The TATA element (box A) has since been observed upstream of most of the isolated archaeal genes (reviewed in [4, 10]). The consensus TATA elements of *Archaea* are summarized in Fig. 2. Although in most archaeal promoter regions TATA elements have been recognized, it should be noted that in a number of archaeal promoters the box A nucleotide sequence as well as its position relative to the transcription start less clearly resembles the described eukaryal-like promoter (see Fig. 2B). This might well correspond to variations in the efficiency of transcription initiation [73].

The transcription signal designated box B is certainly not ubiquitous in archaeal promoters. Based on a relatively limited number of sequences with established transcription start sites it was proposed that a pyrimidine-purine dinucleotide consensus was located 27+4 downstream center of box A, at the transcription start [4, 10].

A breakthrough in the elucidation of the archaeal transcription machinery has been the identification of transcription factors of the Euryarchaeon *Metha-*

Fig. 2 A, B. Comparison of archaeal promoter elements: A consensus sequences of TATA-elements in mapped archaeal promoters of *Sulfolobales* (*Crenarchaeota*), and halophilic and methanogenic *Euryarchaeota* after Thomm [73]; the consensus of thermophilic *Euryarchaeota* (*Thermococcales*) is based on mapped promoters from *Pyrococcus furiosus*, as shown below; B mapped promoters of *Pyrococcus furiosus*: glutamate dehydrogenase (*gdh*) [152], β-glucosidase (*celB*) and alcohol dehydrogenase (*adhA*) [110, 110a], glyceraldehyde:ferredoxin oxidoreductase (*gor*) [153], pyrolysin (*pls*) [147], DNA polymerase (*dpo*) [148], protease I (*pfpI*) [149], pyruvate water dikinase (*pdk*) [150], hydrogenase (*hydB*) [151]. Indicated (bold/underlined) are proposed boxA (TATA), reported transcription start (+1); gaps are introduced to align the purine rich ribosome binding sites (rbs), and ATG/GTG startcodons (ATG). In the *gdh* promoter, directed mutagenesis of boxA (TTTATA changed to TTTAGA) caused a dramatic drop of in vitro transcription [152]. In the case of *gdh*, *celB*, *adhA*, and *gor* the transcription starts have been determined in vivo (primer extension on mRNA) as well as in vitro (primer extension on in vitro synthesized transcripts) [110a, 152, 153]

A.

Sulfolobales (*Crenarchaeota*)	(T/C) – T – T – A –(T/A)– A – T
Halobacteria (*Euryarchaeota*)	(T/C) – T –(T/A)–(A/T)– A
Metanococcales (*Euryarchaeota*)	(T/A)–(T/A)– T – A – T – A – T
Thermococcales (*Euryarchaeota*)	(T/A)– T – T – A – T – A
archaeal consensus	T – T – T – A –(T/A)– A

B.

	TATA	+1	rbs	ATG
gdh	AAGCT**TTATA**TAGGCTATTGCCCAAAAATGTATC**G**--CCAATCACCTAATTT**GGAGGG**ATGAAC**ATG**
celB	AAATA**TTATA**AATCACAATATCAAAATATAAAGCT**A**----------------**GAGGTGG**AAAGT**ATG**
adhA	AAAAA**TTATA**AAAAACATCAAGCTTATATTGCTGG**G**------------------**AGGGA**TAAA**ATG**
gor	AAAAT**TTTTA**AGTAAGTTAAATGAAATCACGTCACT**G**GTAGTGGTATAATC**GAGG**TGATGACGT**ATG**
pls	AATGT**TTATA**ATTGGAACGCAGTGAATATACAAAA**T**-GAATATAACCTC**GGAGGTGA**CTGTAGA**ATG**
dpo	AGGTT**TTATA**CTCCAAACTGAGTTAGTAGAT**A**------------------T**GTGGGGAG**CATA**ATG**
pfpI	ATGCC**TTAAA**GAAAAGCCACGAATAAAGTCTTT**G**--------------------**GTGA**TAGGA**ATG**
pdk	TAATT**TTAAA**TATAGCTCACCTTTATCACTCAC**G**---GTTATTTTAAGGC**GGAGGTGAA**CTGAA**ATG**
hydH	AACGA**AAATT**TGAGGAGTATTGGTCAATTATGC-----------TCATT**GGGAGGTGG**TTTGT**GTG**

---

nococcus thermolithotrophicus as well as the Crenarchaeon *Sulfolobus shibatae*: aTFA and aTFB [74, 75]. Sequence information derived from the corresponding genes, indicated striking homology with the eukaryal TBP (aTFB) and TFIIB (aTFA) [76–80]. Alignment of the archaeal and the eukaryal TBP indicated that the latter has an additional N-terminal domain that appears not to be well conserved; the C-terminal domain is involved in the interaction with the TATA element (see below). Making use of the isolated components (TBP, TFIIB, RNA polymerase and TATA-box containing DNA fragment), cell-free transcription systems have been developed for *Methanococcus thermolithotrophicus*, *Methanobacterium thermoautotrophicum*, *Sulfolobus shibatae*, and *Pyrococcus furiosus*, as recently reviewed in [73]. The apparent homology between the transcription initiation process in *Archaea* and *Eukarya* has been corroborated by the demonstration that the archaeal TFs could be substituted by their respective eukaryal (human, yeast) counterparts in in vitro transcription assays [81, 82].

4.4
Structural Information

Recently, the crystal structure of a eukaryal TFIIB-TBP-TATA-element ternary complex has been solved [83]. This study confirmed the anticipated inter-action of the TATA element and the TATA-binding protein. The recognition of the proximal promoter element (TATAAAAG; –31/–24) by TBP was demon-strated to proceed through an induced-fit mechanism, on the one hand via a subtle conformational changes of the TBP polypeptide, but on the other hand via a rather dramatic distortion of the DNA molecule, a 110 °C angle between the incoming and outgoing double helix [83]. It is tempting to assume that the bending of this "upstream" promoter region either enhances the interaction of the RNA polymerase with the "downstream" promoter region, or contributes to its unwinding (see Fig. 3). A curved anti-parallel β-sheet of TBP interacts with the phosphoribose backbone as well as with the TATAAAAG bases in the minor groove of the double helix. Both TFIIB and TBP have been repor-ted to interact with transcriptional activators and co-activators (reviewed in [83]). The structure revealed that TFIIB not only interacts with TBP, but also with the phosphoribose backbone (not with the bases) of the DNA fragment, both upstream (–38/–30) and downstream (–23/–14) of the TATA element (–31/–24). The N-terminal domain of the human TFIIB, which has been deleted in the aforementioned co-crystallization study, contains a Cys-X_2-Cys-X_{17}-Cys-X_2-Cys motif that resembles a metal-binding site as in Zn-fingers. Apparently, however, the latter domain does not play a role in DNA binding, but rather appears to be involved in an interaction with the RNA poly-merase II/TFIIF complex, and probably with some transcriptional (co)ac-tivators. This is in agreement with the predicted model [83], in which the N-terminus of TFIIB is facing the downstream site of the ternary complex (Fig. 3). Indeed, it has been demonstrated that TFIIB mutants alter the RNA polymerase II start sites in yeast, suggesting that TFIIB functions as a spacer between TBP and the polymerase complex at the core promoter [56]. The C-ter-minal domain of TFIIB, containing two 84-amino acid direct repeats, is essential for interaction with the TFIID(TBP)-TATA-element complex; it does not, how-ever, bind to the TFIID(TBP) in the absence of the TATA-element.

The three-dimensional structure of the N-terminal domain of the *Pyrococcus furiosus* TFIIB, which has high sequence homology with eucaryal analogues, is strikingly similar to that of the "zinc ribbon" of the eucaryal transcription elonga-tion factor TFIIB [84]. In addition, the three-dimensional structure of the TATA-box binding protein (TBP) from *Pyrococcus woesei* has been solved. As expected from sequence homology (36–41% identical to eukaryotic TBPs) its overall struc-ture is very similar to eukaryotic TBPs. Titration calorimetry data show that the affinity of *Pyrococcus* TBP for its DNA target, unlike its eukaryotic counterparts, is enhanced by increasing the temperature and salt concentration [84a]. The recently solved preinitiation complex of P. woesei (TBP, TFIIB, TATA-element) revealed that indeed the overall structure is essentially the same as their eukaryal counterparts [84b] (Fig. 3).

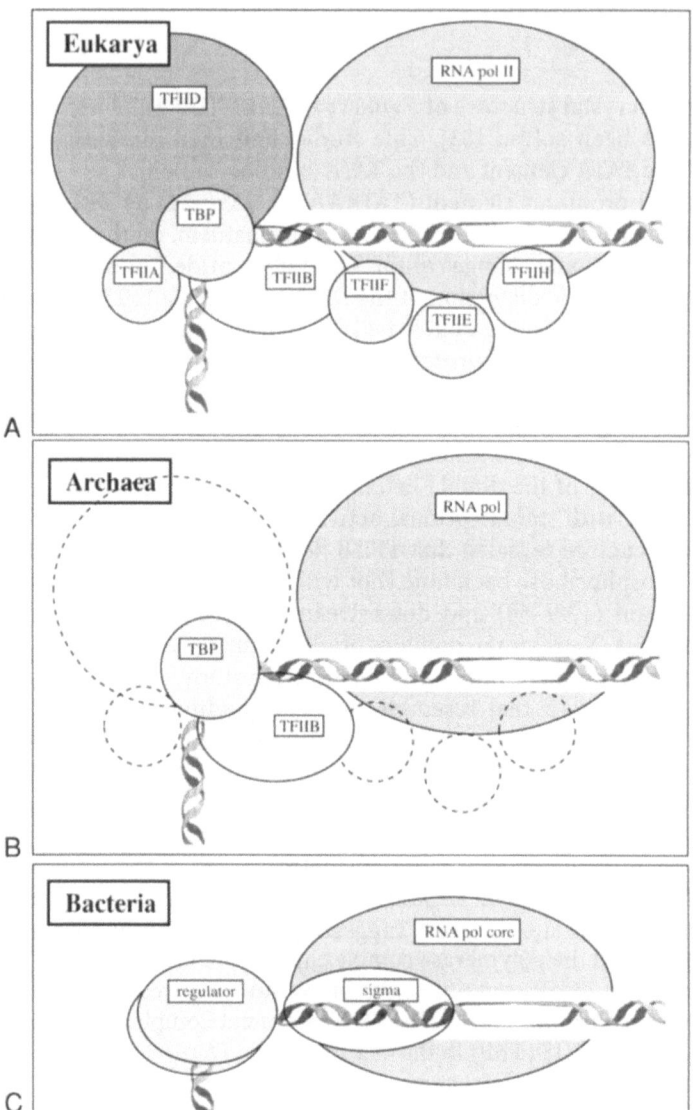

Fig. 3A–C. Cartoon of transcription initiation complexes of: A *Eukarya*; B *Archaea*; C *Bacteria*. The promoter region is represented by a *bar,* the locally unwound DNA helix by a *"split" bar;* transcription factors are drawn as *spheres.* The TBP (A, B) is composed of two duplicated domains; the TFIIB (A, B) has an N-terminal zinc (Zn)-containing domain, and two C-terminal duplicated domains (TFIIB-C1,-C2). Like the DNA bending of the eukaryal promoter brought about by interaction of the transcription factors TBP/TFIIB (A) [83], a bacterial promoter has been reported to be bent as well when complexed with a transcription activator (*lac* promoter complexed with Crp/cAMP-dimer [146]) (C), probably facilitating the RNA polymerase/promoter interaction or the DNA helix unwinding. Although not required in archaeal in vitro transcription systems (see text [73]), it may be that some additional transcription factors are involved in fine-tuning of the archaeal transcription initiation regulation (hypothetical TFs are represented by *dotted lines* although no eukaryal homologs have been identified in the completed archaeal genome sequences [117, 154, 155]). For further discussion, see text

4.5
Transcription Elongation

Genes have recently been isolated from *Sulfolobus acidocaldarius* and *Thermococcus celer* [64, 85], the derived products of which share homology with the C-terminal domain of the eukaryotic transcriptional factor TFIIS, including a putative zinc binding site. Characterization of the eukaryal TFIIS revealed that, in addition to this C-terminal domain, its N-terminal domain is required for binding to the polymerase. Both domains of the eukaryal TFIIS are required for transcription elongation by RNA polymerase II [86]. Interestingly, the archaeal genes also appear to be structurally related to two eukaryal RNA polymerase subunits. Although the gene that codes for the *S. acidocaldarius* TFIIS-like polypeptide is located adjacent to the *rpo*L gene encoding the RNA polymerase subunit L, the former protein apparently does not co-purify with the *Sulfolobus* RNA polymerase [64]. bacterial factors with an analogous role in transcription elongation, e.g. *E. coli gre*A and *gre*B [87], do not share an apparent sequence homology with TFIIS.

4.6
Transcription Termination

Archaeal genes are often located in clusters as on the bacterial chromosome (see Sect. 6 – Gene Organization and Regulation). However, despite this physical linkage, they are not necessarily transcribed as polycistronic messengers. Termination of transcription in *Archaea* appears to be indicated by a T-rich polypyrimidine sequence, ranging from 4 to 30 nucleotides, downstream inverted repeats and immediately upstream the termination site. In many instances, no unique termination site can be determined, as indicated by S1 nuclease mapping of the 3′ terminus of archaeal transcripts. However, since no common pattern of secondary structure is present among archaeal terminator regions, it is unlikely that transcription termination in *Archaea* resembles the bacterial Rho-independent termination.

  Post-transcriptional modification in *Archaea* has been described in a number of instances. Some reports are available on shorter and longer poly(A) tails in archaeal mRNA [88]. Unlike eukaryal mRNA, however, 5′ capping has never been observed in archaeal messengers. The maturation via splicing of transcribed RNA molecules is well established for tRNA and rRNA [89]. Apart from the genes encoding stable RNA, apparently no introns are present in *Archaea*.

5
Translation

During protein synthesis the mRNA nucleotide sequences are translated into (poly)peptide amino acid sequences. This process requires the coordinated interplay of many different components: mRNA, amino acids, tRNA, aminoacyl tRNA synthetase, the ribosome complex that consists of three rRNA molecules and dozens of proteins (50–70), and in addition a set of polypeptide factors that assists in translation initiation, elongation and termination (reviewed in [89]). The translation process is well conserved in all (existing) life forms, and hence analysis of its key components turned out to be a powerful tool in estimating phylogenetic relations [2, 65, 67]. Although the mechanism of protein synthesis is basically the same in all living cells, some typical features have been noticed in comparative studies on the three phylogenetic lineages.

5.1
Bacterial and Eukaryal Translation

In *Bacteria,* the translation of the mRNA often starts before transcription is completed ("co-transcriptional translation"). Moreover, the rate of de novo protein synthesis is further enhanced by the fact that a number of ribosomes simultaneously translate a single mRNA molecule (polysome). The proper in-frame translation of either mono- or polycistronic bacterial mRNA is established via two base pairing events. First, formation of 3–9 base pairs between the purine-rich Shine-Dalgarno sequence (positioned from –13 to –5 upstream the translation start in *E. coli* [90]) and a complementary poly-pyrimidine sequence near the 3′ terminus of the 16 S rRNA component of the small 30 S ribosomal subunit. The efficiency of the latter interaction in *E. coli* is enhanced by elongation factor EF-3. Second, the actual protein synthesis is initiated by the entry of the tRNA-coupled formyl-methionine (fMet-tRNA$_f$) by base pairing of the tRNA$_f$ anticodon with the AUG start codon located 25 bp or more from the 5′ end of the mRNA. The assembly of this initiator tRNA and the mRNA-30 S ribosomal complex is mediated by two alternative translation factors (IF-1 and IF-2 in *E. coli*) that assist in the formation of the 30 S initiation complex. Subsequently, the 50 S ribosomal subunit associates to the initiation complex, and the tranformation process continues with the elongation phase. The role of a number of polypeptides in transformation elongation is well established. An elongation factor EF-Tu mediates the GTP-driven delivery of additional aminoacyl-tRNAs. A peptidyl transferase component of the 50 S ribosomal subunit is involved in the subsequent peptide bond formation. The last phase of an elongation cycle concerns the dissociation of the uncharged tRNA from the P-site, and the movement of the peptidyl-tRNA-mRNA complex from the A-site to the P-site. This is catalyzed by a second G-protein, the elongation factor EF-G (reviewed in [91]).

In *Eukarya* genes are not organized in operon structures. The monocistronic mRNA molecule often undergoes extensive processing: 5′ capping of mRNA (5′–5′ pyrophosphate linkage of 7-methylguanosine), 3′ polyadenylation of mRNA, 3′ CCA tri-nucleotide addition of tRNA, splicing of a 5′ leader sequence of tRNA, and excision of internal non-coding regions of tRNA and rRNA. Recognition of the translation start in *Eukarya* is completely different from the bacterial mechanism. Instead of a Shine-Dalgarno sequence, the 5′ cap of the eukaryal mRNA binds to the ribosome (generally the small subunit) and via a scanning mechanism the AUG start codon is detected. The eukaryal ribosome consists of a 40 S small subunit with 17 S rRNA and about 35 proteins, and a 60 S large subunit composed of 18 S, 5.7 S, 5 S rRNA and about 45 proteins. The mRNA/ribosome association and the subsequent translation initiation involves the concerted action of a larger number of initiation factors (up to ten). Unlike the bacterial situation, the first amino acid of the eukaryal nascent chain is a non-modified methionine [92].

5.2
Archaeal Translation

Compared to bacterial and eukaryal tRNAs, the archaeal homologues contain a couple of unique modified nucleotides (e.g. pseudouridine, 1-methylguanosine) [93]. In addition, the CCA stem of initiator tRNAs appears to contain a distinct stretch of five base pairs, not found in eukaryal and bacterial molecules. Like eukaryal tRNAs, the 3' amino acid attachment site (CCA) in most archaeal tRNAs is not encoded by the corresponding genes, but rather results from a post-transcriptional modification [94]. Recent sequence comparison of the CCA-adding enzyme (ATP(CTP):tRNA nucleotidyl transferase) from *Sulfolobus shibatae* with bacterial and eukaryal homologues has revealed that the enzymes form a superfamily together with the poly(A) polymerases [94a]. Another eukaryal feature of archaeal tRNA is the presence of introns [95].

The enzymes that catalyze the ATP-driven activation of amino acids, amino-acyl-tRNA synthetases, constitute an extensive multigene family (reviewed in [96]). Numerous bacterial and eukaryal aminoacyl-tRNA synthetase sequences have been aligned and phylogenetic trees constructed from them. It was found that analogous synthetases, specific for the same amino acid, from a variety of organisms group together. This is in agreement with the idea that the amino-acyl-tRNA synthetases are very ancient enzymes that evolved long before the divergence leading to prokaryotes and eukaryotes, resulting in specific synthe-tases for the full complement of 20 amino acids. The enzymes that are divided into two distinct groups appear to have evolved from independent roots. Group I contains the enzymes specific for glutamic acid, glutamine, tryptophan, tyro-sine, valine, leucine, isoleucine, methionine, and arginine. Group II enzymes include those activating threonine, proline, serine, lysine, aspartic acid, aspara-gine, histidine, alanine, glycine, and phenylalanine. Since both groups contain a spectrum of amino acid types, it has been proposed that each could have once supported an independent system for protein synthesis. Within each group, enzymes specific for chemically similar amino acids tend to cluster together, indicating that a major theme of synthetase evolution involved the adaptation of binding sites to accommodate related amino acids with subsequent specializa-tion to a single amino acid. In a few cases, however, synthetases activating dissimilar amino acids are grouped together [97].

Rooted phylogenetic trees for both elongation factors and ATPase subunits showed *Bacteria* as branching first from the universal tree with *Archaea* and *Eukarya* as sisters (Fig. 1A) [65, 66]. Recently, a number of gene fragments coding for isoleucyl-tRNA synthetases of several distant species have been iso-lated and compared [67]. The organisms represent deep evolutionary branches of *Eukarya* (*Nosema locustae*), *Bacteria* (*Aquifex pyrophilus* and *Thermotoga maritima*) and *Archaea* (*Pyrococcus furiosus* and *Sulfolobus acidocaldarius*). All isoleucyl-tRNA synthetase trees provided strong support for a monophylic origin of *Crenarchaeota* and *Euryarchaeota*. Moreover, the presented trees all indicated that *Archaea* and *Eukarya* are sister groups, confirming previously reported tree rooting [65, 66]. One should be cautious, however, to refer to these data as "the universal tree", because of the conflicting results obtained by phylogenetic analysis of different classes of polypeptides (see discussion above, Fig. 1) [4, 68].

Archaeal ribosomes resemble their bacterial counterparts with respect to the rRNA components (23 S, 16 S and 5 S). Although the archaeal ribosome apparently is composed of more polypeptide components than the bacterial ribosome (70 rather than 50), the sedimentation behaviour of ribosomes from both lineages (70 S), as well as that of their dissociated subunits (30 S, 50 S), is similar [88]. The genes encoding the 16S-23 S rRNA genes are organized as an operon. As in *Bacteria*, but unlike the situation in some *Eukarya*,the small 5 S rRNA gene is located elsewhere on the chromosome. Although direct experimental data on base pairing between archaeal mRNA leaders and the complementing 3' end of 16 S rRNA is not available, purine-rich sequences have been noticed to preceed the initiation codon of mRNAs from thermophilic and methanogenic *Euryarchaeota* (reviewed in [10,88]). A Shine-Dalgarno (SD)-like function of these archaeal motifs in translation initiation has been corroborated by a large number of successful heterologous expression studies of archaeal genes in *E. coli*. In halophiles, however, putative SD motifs often appear to be located 1-4 base pairs downstream the start codon [88]. Together with the presence of very short (or lacking) leader sequences, it has been suggested that the actual mRNA/ribosome interaction, at least in some halophilic *Archaea*, is structurally different from the bacterial-like situation. In *Crenarchaeota* like *Sulfolobus*, putative SD motifs have been noticed in a number of instances, but certainly not upstream every gene. Synthesis of most archaeal proteins appears to start at AUG codons (AUG 82%, GUG 15%, UUG 3%) [10]. The AUG codon, often preceded by an SD-like sequence, determines the translation start. This codon interacts with a Met-tRNA$_m$, similar to the eukaryal situation [4].

A putative homologue of the eukaryal translation initiation factor 5 A, eIF-5 A, has been isolated from *Sulfolobus acidocaldarius* [98]. The exact role of eukaryal eIF-5 A and its archaeal counterpart in the translation process, however, remains to be solved. A second archaebacterial IF homologue has recently been reported, eIF-1 A [99]. The eukaryal eIF-1 A polypeptide plays a role in translation initiation, analogous to the bacterial IF-3: mediating the binding of mRNA to the small 40 S ribosome subunit, and preventing the 40 S and 60 S subunits from associating in the absence of mRNA [100].

Polypeptides assisting in the archaeal translation elongation process (the elongation factors EF-1α, EF-2) have been useful in the construction of a rooted phylogenetic tree [65]. The latter study strongly suggested that the archaeal and eukaryal elongation factors are clustering, and that the bacterial homologs (EF-Tu, EF-G) are more distantly related (see above). *Methanococcus vannielii* displays an *rps12-rps7-fus-tuf* gene organization, encoding two ribosomal proteins S12 and S7, as well as two translation elongation factors EF-1α and EF-2, resembling the *E. coli* streptomycin resistance *(str)* operon (reviewed in [88]). On the other hand, the gene encoding elongation factor EF-1α in *Pyrococcus woesei* and *S. acidocaldarius* is part of a gene cluster with the genes coding for ribosomal protein S10 and tRNA^Ser [101]. In the latter organisms, and in *Desulfurococcus mobilis*, the gene of EF-1α does not cluster with that of EF-2 [102].

6
Gene Organization and Regulation

Archaeal genomes contain clusters of cooperative genes that often resemble counterparts in *Bacteria*, both with respect to gene composition and gene order. The archaeal gene clusters, however, differ in some aspects from the bacterial operons.

6.1
Gene Clusters

A number of the archaeal gene clusters have been reported (i) to contain internal promoters and terminators, and (ii) to produce transcripts of varying lengths, both poly- and mono-cistronic [89]. A well-characterized example concerns a genomic cluster of *Sulfolobus acidocaldarius* with the *rpo*B, *rpo*A', and *rpo*A" genes [64]. The three RNA polymerase subunit genes are transcriptionally linked, like in *Bacteria*. However, shorter transcripts encoding only *rpo*A" and some ribosomal protein (S12), the gene of which is located downstream *rpo*A", are produced along with the *rpo*BA'A"-polycistronic mRNA. It is not clear whether the apparent differential transcription arises from different transcription initiation or from post-transcriptional processing event [103].

Despite the interest recently devoted to molecular evolution and genetics in hyperthermophiles, the regulation of gene expression in these organisms has not yet been investigated in great detail. It has been clearly established, however, that the archaeal basal transcription machinery is strikingly similar to the eukaryotic one (see Sect 4 – Transcription). On the other hand, in several cases the organization of analogous genes is colinear in *Archaea* and *Bacteria*. In both lineages functionally related genes are often cotranscribed or physically linked.

The organization of ribosomal RNA operons in *Archaea* has been extensively studied [104, 105]. In all cases studied, archaeal 16 S and 23 S rRNA genes are linked. Unlike some eukaryal rRNA gene clusters, the short spacer between these rRNA genes does not contain a tRNA gene. The 5 S rRNA is usually not linked to the larger rRNA cluster. The 16S–23 S rRNA cluster is flanked by inverted repeats that probably play a role in processing. In a few species introns have been found, either in the 16 S or in the 23 S rRNA (see Sect. 3 – Mobile Elements). Whereas in *Bacteria* and *Eukarya* many copies of the rRNA gene clusters have been reported (*E. coli* 7 copies, *Saccharomyces cerevisiae* over 100 copies), thermophilic *Archaea* appear to contain only a single copy [104, 105]. The genes coding for ribosomal proteins are clustered in operons with the same organization found in *Bacteria* [106], although the total number of ribosomal proteins in *Archaea* is significantly higher (see Sect. 5 – Translation). Indeed, several genes that appear to be absent in *Bacteria* are located either within or adjacent to the archaeal operons. Some of them encode homologues found in the eukaryal ribosome, other appear to have no equivalents in the other lineages.

In *S. solfataricus*, the genes *trpE*, *trpG* and *trpC*, coding for enzymes of the tryptophan biosynthetic pathway, are linked and are presumably cotranscribed. However, their relative order is different from that found both in *Bacteria* and in other *Archaea* [107]. The genes encoding the four subunits of the *Pyrococcus furiosus* sulfhydrogenase, a chimeric (Ni-Fe) hydrogenase/sulfur reductase complex, are organized as a gene cluster (*hydBGDA*). Although no transcription analysis has been performed, the structural features of the *hyd* locus suggest that, like the multimeric hydrogenases of, for instance, *E. coli*, the enzyme is encoded by a single transcription unit. First, the genes are tightly linked: there are overlapping genes coding for *hydB and hydG*, as well as for *hydD* and *hydA*; the genes *hydG* and *hydD* are separated by ten nucleotides. Second, two transcription termination signals are located immediately downstream the last gene, but not at any other position of the gene cluster. No canonical transcription start site was found, although A/T rich streches are located upstream of the first gene [108].

A tight gene organization is not always indicative of co-transcription, as suggested by the studies on the locus *lacS*/orf2 of *S. solfataricus* [109]. Operons for carbohydrate metabolism have been used as the paradigm of transcriptional regulation in *Bacteria,* and they include genes for structural, regulatory and transport proteins. *The lacS* gene encodes a β-glycosidase in *Sulfolobus solfataricus* and orf2, mapping immediately upstream of it, encodes a putative permease. The two genes are only 20 bp apart, and no potential archaeal terminator sequence has been found in the intergenic region. However, they are transcribed as monocistronic RNAs from independent promoters. Both transcripts show a peak of expression at the exponential phase of growth, but their relative abundance is very different: the *lacS* messenger is at least ten-fold more abundant than that of orf2 [109]. The homologue of *lacS* in *Pyrococcus furiosus, celB*, is also transcribed as a single unit, and the organization of the locus is different from that of *S. solfataricus* [110]. The pyrococcal *celB* gene is preceded by a gene cluster, transcribed in the opposite orientation, that encodes two types of alcohol dehydrogenase, an endoglucanase and a biotin ligase-homologue. The four genes constitute an operon as has been demonstrated by the fact that they are transcribed as a monocistronic messenger [110a].

6.2
Regulation

The first example of transcriptional regulation described in hyperthermophiles was the UV-inducible transcript of the virus SSV1 of *S. shibatae* [111], which presumably is involved in the viral response to UV irradiation (see Sect. 3 – Mobile Elements). More recently, the cluster encoding the 16S and 23S rRNA genes from *Pyrococcus furiosus* has been studied. In response to a 20-fold increase in growth rate transcription of 16 S rRNA is up-regulated 7.5-fold. [10, 112]. Several co-regulated genes or transcriptional units from different strains have been described, but in no case have regulatory circuits of gene expression been clarified.

One of the gene activation pathways extensively studied in both Prokaryotes and Eukaryotes is the induction of heat-shock genes. The same phenomenon also occurs in hyperthermophilic *Archaea*. Upon exposure to temperatures of 5 – 10 °C higher than the growth temperature (85 – 90 °C), *S. shibatae* induces the expression of a high molecular mass complex, "rosettasome", which consists of equal proportions of subunits α and β (the latter previously called TF55) [113]. The rosettasome resembles the bacterial HSP60 complexes both in structure and chaperonin activity [114, 115]. On the other hand, the sequence of the two archaeal subunits shows significant similarity with the eukaryotic chaperonins TCP1. The synthesis of the α and β subunits, which are among the most abundant proteins at normal temperature, significantly increases during heat-shock, and they are the only two proteins synthesized above 90 °C [113]. This coordinated regulation occurs at transcriptional level rather than different stability. The genes encoding the two subunits are not linked; the search for regulatory sequences in their upstream regions revealed no similarity to those found in bacterial or eukaryotic heat shock genes. Distinctive features common to the upstream regions of the α and β subunits encoding genes are a pseudo-palindromic sequence, with the features of a binding site for a transcription factor. In addition, an extended TATA element (box A) is present in the promoter region, also found in other highly expressed non-heat shock genes. The role of these sequences in the regulation of heat-shock genes has not been analyzed experimentally.

Trans- or cis-acting elements involved in regulation of archaeal gene expression have not yet been characterized in great detail. Recently, the genes of a limited number of archaeal transcription regulators have been reported. First, a gene encoding a putative homologue of the leucine-responsive regulatory protein (Lrp) family has been identified downstream of the glutamate dehydrogenase gene of *Pyrococcus furiosus* [116, 116a]. Homologues have also been identified in the genome sequences of *Methanococcus jannaschii* (117), and *Sulfolobus solfataricus* (117a). Lrp is a global transcriptional regulator in *Escherichia coli* and related *Bacteria*, either stimulating or repressing the expression of a large number of genes and operons [118]. Interestingly, the genome analysis of *Pyrobaculum aerophilum* revealed the presence of a homologue of the transcription repressor of the *E. coli* lactose operon (LacI), as well as a homologue of a sigma factor (NtrA) of *Azorhizobium* sp. [119].

Although similar DNA binding helix-turn-helix motifs are certainly present in a number of eukaryal regulators, (e.g. the homeo domains; reviewed in [119a]), no sequences with significant overall homology to transcription factors and sigma factors from *Bacteria* have yet been identified in *Eukarya*. Considering the discussed eukaryal features of the archaeal transcription machinery, the aforementioned finding of homologues of bacterial transcription factors strongly suggests the interesting possibility of the coexistence of bacterial- and eukaryal-like mechanisms controlling gene expression in *Archaea*.

7
Cloning and Heterologous Expression

The analysis of genes and the corresponding polypeptides from hyperthermophilic *Archaea* relies on the functional expression in heterologous systems,
either bacterial (*E. coli, B. subtilis*) or eukaryal (*S. cerevisiae*) [120]. Many examples are available of efficient expression and recovery of functional thermophilic enzymes in *E. coli* [24, 121]. This demonstrates that the correct folding can
often be obtained in enviroments that are very different from the native conditions and that the thermal stability rules of many proteins from hyperthermophilic origin are dictated by the amino acid sequence. Nevertheless, the
functional production of some enzymes was not successful in mesophilic bacterial hosts. This can have several reasons, e. g. archaeal expression signals, biased
codon usage, requirement of specific polypeptides involved in processing or
assembly. In addition, in some cases thermostability of polypeptides requires
the presence of extrinsic stabilizing factors (compatible solutes [121a]). Strategies to overcome these problems of functional heterologous production are
codon modification [122], or the use of eukaryal expression systems, although
the examples reported until now are limited to yeast and insect cells [123–125].
However, the folding of certain proteins from thermophiles may require elevated temperatures, high ionic strength [124] or specific molecular chaperones.

Although much of our knowledge on polypeptides from hyperthermophiles,
and of the corresponding gene organization described in the previous sections
has been acquired by analysis of heterologous expression, an obvious limitation
of this approach concerns the fact that proteins requiring post-translational
modification(s), e. g. processing of signal sequence, glycosylation, incorporation
of an exotic co-factor, will not be synthesized in their active configuration. This
is the motivation to develop multicopy expression vectors for thermophilic
Bacteria and *Archaea,* to enable, for example, directed mutagenesis and homologous expression (see below).

The comparison of amino acid sequences of thermophilic and mesophilic
enzymes has been used to investigate the molecular basis of thermal stability of
a number of polypeptides [126]. Although many attempts have been made, no
solid general rules on protein stability could be deduced from sequence alignment analysis. The first convincing intrinsic stabilizing features were the result
of comparing high resolution three-dimensional structures. Several 3-D structures obtained by either crystal X-ray diffraction or NMR analysis of proteins
from hyperthermophilic *Archaea* have recently been reported [127–133a]. The
main conclusion derived from a comparison with a mesophilic homologue is
that protein hyperthermostability is achieved by minimal alterations of the
mesophilic fold, and with only a few extra stabilizing interactions. Such interactions are mainly electrostatic: hydrogen bonds and salt bridges (reviewed in
[8]). In the case of the glutamate dehydrogenase (GDH) the salt bridges are or-
ganised in large networks involving groups at subunit or domain interfaces [6,
133]. In glyceraldehyde-3-phosphate dehydrogenase, salt bridges appear to be
on the polypeptide surface, stabilizing exposed domains [7]. Comparison of the
crystal structure of the *Pyrococcus woesei* TATA-binding protein and its meso-

philic eukaryal homologue revealed several differences: a disulfide bond not found in mesophilic counterparts, an increased number of surface electrostatic interactions, and a more compact protein packing [84a]. In a structural modelling study in which several serine proteases from mesophiles, thermophiles and hyperthermophiles have been compared, aromatic-aromatic interaction at the polypeptide surface has been proposed as an additional stabilizing feature [133b].

From an engineering point of view, one should realize that there appear to be no general rules by which a polypeptide acquires a higher thermal stability, and hence that a directed engineering approach will be extremely difficult.

8
Development of Genetic Systems

Whereas molecular genetics techniques have been fully developed for halophiles (reviewed in [134]), and partly for methanogens [135], appropriate tools for genetic manipulation of hyperthermophilic *Archaea* are still lacking. The absence of a system that would allow the characterization of gene function in vivo and, in more general terms, for the dissection of fundamental cellular processes, is a serious limitation. With the completion of several genome projects of hyperthermophilc *Archaea* expected shortly, revealing many uncharacterized "reading frames", the development of specific tools and techniques becomes even more urgent.

Essential tools to this aim are suitable selective markers, vector systems and tranformation techniques. As to selective markers, detailed studies on the resistance of different hyperthermophiles to antibiotics and other drugs have been reported [136–139]. More recently, a spontaneous point mutation has been identified in the 23 S rRNA gene of *Sulfolobus acidocaldarius*, which confers resistance to chloramphenicol, carbomycin and celesticetin [140]. In addition, a 23 S rRNA mutant of *Pyrococcus furiosus* with resistance to thiostrepton has been isolated [141]. The β-glycosidase/galactosidase encoding genes, like *lac* S of *Sulfolobus solfataricus* [142] and *cel*B of *Pyrococcus furiosus* [110] are promising reporter genes; indeed, *lac* S has been recently used as a marker for the screening of transformants (see below).

Considerable effort has recently been devoted to the isolation of transducing phages and autonomously replicating plasmids from hyperthermophic *Archaea*. The SSV1 virus is a natural lysogen of *Sulfolobus shibatae*, not found in the *Sulfolobus solfataricus* genome. It has been shown that the genome of SSV1 can be transmitted to *Sulfolobus solfataricus* either by infection or by electroporation; in this latter case, exogenous DNA can be incorporated in the viral genome [143]. Few autonomously-replicating plasmids have been isolated; among them, the most promising for the construction of cloning vectors are pGT5 from *Pyrococcus abyssi* pGE5 [144] and pNOB8 from a *Sulfolobus* strain isolated recenty [48]. The former is particularly suitable since it is small (3.5 kb) and has been completely sequenced. In contrast, plasmid pNOB8 is large (45 kb) and neither the genetic map nor its sequence are yet available; both its size and its high copy number probably affect cell growth. The pNOB8 vector is efficiently transmitted from the original isolate to different *Sulfolobus* strains; the transfer does not involve the release of virus-like particles, but occurs through cell-cell interactions, with a mechanisms resembling bacterial conjugation [48].

A recombinant derivative of pNOB8 has been used to transform a *Sulfolobus* strain [145]. The gene *lac* S, encoding the β-glycosidase/galactosidase of *Sulfolobus solfataricus*, under the control of the strong promoter of the ribosomal protein S12, was inserted in a non-essential site of pNOB8. By electroporation, this recombinant plasmid has been successfully used to transform a *S. solfataricus* strain containing an insertion sequence in *lac* S. Subsequently, β-galactosidase positive colonies were recovered efficiently, since after the first event of transformation, the plasmid spreads over the entire culture through conjugation. The plasmid is maintained in the episomal state in the recipients, but after repeated cycles of growth, stable tranformants are selected containing the intact *lac* S gene at the chromosomal locus, most probably due to a recombination event. However, problems of this system are the high copy number of the recombinant plasmid, which results in plasmid instability, and in growth retardation of the transformants.

Recently, the development of an *E. coli-Sulfolobus* shuttle vector has been reported, in which the replication sequence of the SSV1 viral particle (see Sect. 3 – Mobile Elements) and a randomly modified hygromycin marker (selected for thermostability) were inserted in a common *E. coli* vector [145 a]. In addition, the construction of a pGT5/pUC18-derived shuttle vector has been described, suitable for cloning and expression of genes in both *E. coli* and *Pyrococcus*. Currently, several potential markers are being included, a.o. point mutants of the 23 S rRNA that give rise to resistance to antibiotics like chloramphenicol and thiostrepton [141], and the alcohol dehydrogenase from *Sulfolobus solfataricus*, allowing growth at elevated alcohol concentrations [145b].

9
Concluding Remarks

Many features of the transcription/translation machinery of *Archaea* are shared either with *Bacteria* or with *Eukarya*. A number of archaeal genes have been demonstrated to be organized in bacterial-like operons, resulting in both mono- and poly-cistronic mRNA. In addition, archaeal mRNAs do not contain a 5′ cap, and only few have been reported to contain a 3′ poly-A strech. During translation initiation, the interaction of the mRNA and the ribosome is also reminiscent of bacteria: a Shine-Dalgarno-like sequence is present just upstream the translation start on the mRNA, and a complementary poly-pyrimidine motif on the 3′ end of the 16 S rRNA. On the other hand, subsequent stages of archaeal translation in many respects resemble the eukaryal processes: the N-terminal methionine is non-formylated, and many rRNA and polypeptide components, e.g. translation factors, and aminoacyl-tRNA synthetases, share highest homology with eukaryal counterparts.

The transcription machinery of distinct types of *Archaea*, in addition, is more akin the eukaryal system. Transcription signals like the RNA polymerase-binding site closely resemble the TATA element of the eukaryal polymerase II. It is anticipated that, besides the recently isolated archaeal homologues of the transcription factors TBP, TFIIB, and probably TFIIS, the identification of more archaeal transcription factors may be expected in the near future. This does not only include proteins that interact with the discussed upstream promoter elements (TATA box), but possibly also transcription factors that interact with more distantly located enhancer-like elements. The unidentified archaeal transcription regulators will be either unique, or related to bacterial and eukaryal regulators with characteristic DNA-binding motifs (e.g. helix-turn-helix, zinc-finger, leucine-zipper).

Because several archaeal genome sequences have recently been completed [117, 154, 155], and even more are to be expected shortly, the research of *Archaea* in general and hyperthermophiles in particular will enter a new phase, with many exciting discoveries to be expected. This will not only help to reveal fundamental matters like many details concerning the evolution and adaptation of extremophiles, but also lead to the discovery of many enzymes with unique properties.

Acknowledgements. Part of this work was supported by the European Union, as part of the BIO-TECH programme BIO2-CT-930274.

10
References

1. Stanier RY, Van Niel CB (1962) Arch Microbiol 42:17
2. Woese CR, Fox GE (1977) Proc Natl Acad Sci USA 74:5088
3. Woese CR, Kandler O, Wheelis ML (1990) Proc Natl Acad Sci USA 87:4576
4. Zillig W, Palm P, Klenk HP, Langer D, Hudepohl U, Hain J, Lanzendorfer M, Holz I (1993) Transcription in *Archaea*. In: Kates M, Kushner DJ, Matheson AT (eds) The biochemistry of *Archaea* (Archaebacteria). Elsevier, Amsterdam London New York Tokyo, p 367
5. Stetter KO, Fiala G, Huber G, Segerer A (1990) FEMS Microbiol Rev 75:117
6. Rice DW, Yip KSP, Stillman TJ, Britton KL, Fuentes A, Connerton I, Pasquo A, Scandurra R, Engel PC (1996) FEMS Microbiol Rev 18:105
7. Jaenicke R (1996) FEMS Microbiol Rev 18:215
8. Vieille C, Zeikus JG (1996) Trends Biotechnol 14:183
9. Forterre P, Bergerat A, Lopez-Garcia P (1996) FEMS Microbiol Rev 18:237
10. Dalgaard JZ, Garrett R (1993) Archaeal hyperthermophile genes. In: Kates M, Kushner DJ, Matheson AT (eds) The biochemistry of *Archaea* (Archaebacteria). Elsevier, Amsterdam London New York Tokyo, p 535
11. Grayling RA, Sandman K, Reeve JN (1996) FEMS Microbiol Rev 18:203
12. Forterre P, Charbonnie F, Marguet E, Harper F, Henckes G (1992) Chromosome structure and DNA topology in extremely thermophilic Archae*Bacteria*. In: Danson MJ, Hough DW, Lunt GG (eds) The Archaebacteria: biochemistry and biotechnology. The Biochemical Society, London, p 99
13. Charbonnier F, Forterre P (1994) J Bacteriol 176:1251
14. Kikuchi A, Asai K (1984) Nature 309:677
15. Forterre P, Mirabeau G, Jaxel C, Nadal M, Daniel M (1985) EMBO J 4:2123
16. Boutier de la Tour C, Portemer C, Nadal M, Stetter KO, Forterre P, Duguet M (1990) J Bacteriol 172:6803
17. Collin RG, Morgan HW, Musgrave DR, Daniel RM (1988) FEMS Microbiol Lett 55:235
18. Bouthier de la Tour C, Portemer C, Huber R, Duguet M, Forterre P (1991) J Bacteriol 173:3921

19. Nadal M, Jaxel C, Portemer C, Forterre P, Mirabeau G, Duguet M (1988) Biochemistry 27:9102
20. Depew RE, Liv LF, Wang JC (1978) J Biol Chem 253:511
21. Jaxel C, Nadal M, Mirambeau G, Forterre P, Takahashi M, Duguet M (1989) EMBO J 8:3135
22. Confalonieri F, Elie C, Nadal M, de La Tour C, Forterre P, Duguet M (1993) Proc Natl Acad Sci USA 90:4753
23. Wallis JW, Chrebet G, Brodsky G, Rolfe M, Rothstein R (1989) Cell 58:409
24. Nadal M, Mirabeau G, Forterre P, Reiter W-D, Duguet M (1986) Nature 321:256
25. Slezarev A, Zaitzek D, Kopylov V, Stetter KO, Kozyavkin S (1991) J Biol Chem 266:12321
26. Bergerat A, Gadelle D, Forterre P (1994) J Biol Chem 269:27663
27. Klimczak LJ, Grummt F, Burger KJ (1986) Biochemistry 25:4850
28. Hamal A, Forterre P, Elie C (1990) Eur J Biochem 190:517
29. Lundberg KS, Shoemaker DD, Adams MW, Short JM, Sorge JA, Mathur EJ (1991) Gene 108:1
30. Kong H, Kucera RB, Jack WE (1993) J Biol Chem 268:1965
31. Pisani FM, Rossi M (1994) J Biol Chem 269:7887
32. Mattila P, Korpela JTT, Pitkanen K (1991) Nucl Acids Res 19:4967
33. Pisani FM, Manco G, Carratore V, Rossi M (1996) Biochemistry (in press)
34. Rossi M, Rella R, Pensa M, Bartolucci S, De Rosa M, Gambacorta A, Gaia CA, Dell'Aversano A, Orabona N (1986) System Appl Microbiol 7:337
35. Elie C, De Recondo AM, Forterre P (1989) Eur J Biochem 178:619
36. Perler FB, Kumar S, Kong H (1996) Adv Prot Chem (in press)
37. Perler FB (1992) Proc Natl Acad Sci USA 89:5577
38. Pisani FM, De Martino C, Rossi M (1992) Nucl Acids Res 20:2711
39. Uemori T, Ishino Y, Toh H, Asada K, Kato I (1993) Nucl Acids Res 21:259
40. Braithwaite DK, Ito J (1993) Nucl Acids Res 21:787
41. Hodges RA, Perler FB, Noren CJ, Jack WE (1992) Nucl Acids Res 20:6153
42. Prangishvili D, Klenk HP (1994) System Appl Microbiol 16:665
43. Fiala G, Stetter KO (1986) Arch Microbiol 145:56
44. Zillig W, Prangishvilli D, Schleper C, Elferink M, Holz I, Albers S, Janekovic D, Götz D (1996) FEMS Microbiol Rev 18:225
45. Palm P, Schleper C, Grammp B, Yeats S, McWilliam P, Reiter WD, Zillig W (1991) Virology 185:242
46. Zillig W, Yeats S, Holz I, Bock A, Gropp F, Rettenberger M, Lutz S (1985) Nature 313:789
47. Zillig W, Kletzin A, Schleper C, Holz I, Janekovic D, Hain J, Lanzendorfer M, Kristjansson JK (1994) System Appl Microbiol 16:609; Keeling PJ, Klenk HP, Singh RK, Feeley O, Schleper C, Zillig W, Doolittle F, Sensen CW (1997) Plasmid (in press)
48. Schleper C, Holz I, Janekovic D, Murphy J, Zillig W (1995) J Bacteriol 177:4417
49. Erauso G, Marsin S, Benbouzid-Rollet N, Bacuher MF, Barbeyron T, Zivanovic Y, Prieur D, Forterre P (1996) J Bacteriol 178:3232
50. Schleper C, Roeder R, Singer T, Zillig W (1994) Mol Gen Genet 243:91
51. Aagaard C, Dalgaard J, Garrett RA (1995) Proc Natl Acad Sci 92:12285, a Grogan DW (1996) J Bacteriol 178:3207
52. Haldenwang WG (1995) Microbiol Rev 59:1
52a. Kunst et al. (1997) Nature 390, 249
53. Collado-Vides J, Magasanik B, Gralla JD (1991) Microbiol Rev 55:3710
54. Riftina F, DeFalco E, Krakow JS (1990) Biochemistry 29:4440
55. Marschall C, Hengge-Aronis R (1995) Mol Microbiol 18:175
56. Parvin J, Sharp P (1993) Cell 73:533
57. Zawel L, Kumar K, Reinberg D (1995) Genes Dev 9:1479
58. Koleske A, Young R (1995) Trends Biochem Sci 20:113
59. Chao DM, Gadbois EL, Murray PJ, Anderson SF, Sonu MS, Parvin JD, Young RA (1996) Nature 380:82
59a. Kornberg RD (1996) Trends Biochem Sci 21:325

60. Zillig W, Stetter KO, Tobien M (1978) Eur J Biochem 91:193
61. Zillig W, Stetter KO, Janekovic D (1979) Eur J Biochem 96:597
62. Leffers H, Gropp F, Lottspeich F, Zillig W, Garrett RA (1989) J Mol Biol 206:1
63. Pühler G, Leffers H, Gropp F, Palm P, Klenk HP, Lottspeich F, Garrett RA, Zillig W (1989) Proc Natl Acad Sci USA 86:4569
64. Langer D, Hain J, Thuriaux P, Zillig W (1995) Proc Natl Acad Sci USA 92:5768
65. Iwabe N, Kuma K, Hasegawa M, Osawa S, Miyata T (1989) Proc Natl Acad Sci USA 86:9355
66. Gogarten JP, Kibak H, Dittrich P, Taiz L, Bowman EJ, Bowman BJ, Manolson MF, Poole RJ, Date T, Oshima T, Konishi J, Denda K, Yoshida M (1989) Proc Natl Acad Sci USA 86:6661
67. Brown JR, Doolittle WF (1995) Proc Natl Acad Sci USA 92:2441
68. Gupta RS, Golding GB (1996) Trends Biochem Sci 21:166
69. Margulis L (1970) Origin of Eukaryotic Cells. Yale University Press, New Haven
70. Thomm M, Wich G (1988) Nucl Acids Res 16:151
71. Reiter WD, Palm P, Zillig W (1988) Nucl Acids Res 16:1
72. Reiter WD, Hüdepohl U, Zillig W (1990) Proc Natl Acad Sci USA 87:9509
73. Thomm M (1996) FEMS Microbiol Rev 18:159
74. Frey G, Thomm M, Brüdigam, Gohl HP, Hausner W (1990) Nucl Acids Res 18:1361
75. Hüdepohl U, Reiter WD, Zillig W (1990) Proc Natl Acad Sci USA 87:5851
76. Rowlands T, Baumann P, Jackson SP (1994) Science 264:1326
77. Marsh TL, Reich CI, Whitelock RB, Olsen GJ (1994) Proc Natl Acad Sci USA 91:4180
78. Ouzounis C, Sander C (1992) Cell 71:189
79. Creti R, Londei P, Cammarano P (1993) Nucl Acids Res 21:2949
80. Qureshi SA, Khoo B, Baumann P, Jackson SP (1995) Proc Natl Acad Sci USA 92:6077
81. Wettach J, Gohl HP, Tschochner H, Thomm M (1995) Proc Natl Acad Sci USA 92:472
82. Gohl HP, Gröndahl B, Thomm M (1995) Nucl Acids Res 23:3837
83. Nikolov DB, Chen H, Halay ED, Usheva AA, Hisatake K, Lee DK, Roeder RG, Burley SK (1995) Nature 377:119
84. Zhu W, Zeng Q, Colangelo CM, Lewis M, Summers F, Scott RA (1996) Nat Struct Biol 3:122,
84a. DeDecker BS, O'Brien R, Fleming PJ, Geiger JH, Jackson SP, Sigler PB (1996) J Mol Biol 264:1072
84b. Kosa PF, Ghosh G, DeDecker BS, Sigler PB (1997) Proc Natl Acad Sci USA 94:6042
85. Kaine BP, Mehr IJ, Woese CR (1994) Proc Natl Acad Sci USA 91:3854
86. Agarwal K, Baek KH, Jeon CJ, Miyamoto K, Ueno A, Yoon HS (1991) Biochemistry 30:7842
87. Sparkowski J, Das A (1990) Nucl Acid Res 18:6443
88. Ramirez C, Köpke AKE, Yang DC, Boeckh T, Matheson AT (1993) Transcription in *Archaea*. In: Kates M, Kushner DJ, Matheson AT (eds) The biochemistry of *Archaea* (Archaebacteria). Elsevier, Amsterdam London New York Tokyo, p 439
89. Amils R, Cammarano P, Londei P (1993) Transcription in *Archaea*. In: Kates M, Kushner DJ, Matheson AT (eds) The biochemistry of *Archaea* (Archaebacteria). Elsevier, Amsterdam London New York Tokyo, p 393
90. Gold L (1988) Ann Rev Biochem 57:199
91. Kurland CG (1992) Annu Rev Genet 26:29
92. Kozak M (1983) Microbiol Rev 47:1
93. McCloskey JA (1986) Syst Appl Microbiol 7:246
94. Kuchino Y, Ihara M, Yabusaki Y, Nishimura S (1982) Nature 298:684
94a. Yue D, Maizels N, Weiner AM (1996) RNA 2:895
95. Kaine BP, Gupta R, Woese CR (1983) Proc Natl Acad Sci USA 80:3309
96. Carter CW (1993) Ann Rev Biochem 62:715
97. Nagel GM, Doolittle RF (1991) Proc Natl Acad Sci USA 88:8121
98. Bartig D, Lemkemeier K, Frank J, Lottspeich F, Klink F (1992) Eur J Biochem 204:751–8

99. Keeling PJ, Doolittle WF (1995) Mol Microbiol 17:399
100. Thomas A, Goumans H, Voorma HO, Benne R (1980) Eur J Biochem 107:39
101. Creti R, Citarella F, Tiboni O, Sanangelantoni A, Palm P, Cammarano P (1991) J Mol Evol 33:332
102. Creti R, Sterpetti P, Bocchetta M, Ceccarelli E, Cammarano P (1995) FEMS Microbiol Lett 126:85
103. Pühler G, Lottspeich F, Zillig W (1989) Nucl Acids Res 17:45171
104. Leffers H, Kjems J, Ostergaard L, Larsen N, Garrett RA (1987) J Mol Biol 195:43
105. Garrett RA, Dalgaard J, Larsen N, Kjems J, Mankin AS (1991) Trends Biochem Sci 16:22
106. Matheson AT (1992) Structure, function and evolution of the archaeal ribosome. In: Danson MJ, Hough DW, and Lunt GG (eds) The Archaebacteria:biochemistry and biotechnology. The Biochemical Society Symposium, London, p 89
107. Tutino ML, Scarano G, Marino G, Sannia G, Cubellis MV (1993) J Bacteriol 175:299
108. Pedroni P, Della Volpe A, Galli G, Mura GM, Pratesi C, Grandi G (1995) Microbiology 141:449
109. Prisco A, Moracci M, Rossi M, Ciaramella M (1995) J Bacteriol 177:1614
110. Voorhorst WGB, Eggen RIL, Luesink EJ, De Vos WM (1995) J Bacteriol 177:7105
110a. Voorhorst WGB, Gueguen Y, Schut G, Dahlke I, Thomm M, Van der Oost J, De Vos WM (1997) (submitted)
111. Reiter WD, Palm P, Yeats S, Zillig W (1987) Mol Gen Genet 209:270
112. DiRuggiero J, Achenbach LA, Brown SH, Kelly RM, Robb FT (1993) FEMS Microbiol Lett 111:159
113. Kagawa HK, Osipiuk J, Maltsev N, Overbeek R, Quaite-Randall E, Joachimiak A, Trent JD (1995) J Mol Biol 253:712
114. Trent JD, Nimmesgern E, Wall JS, Hartl FU, Horwich AL (1991) Nature 354:490
115. Guagliardi AM, Cerchia L, Bartolucci S, Rossi M (1994) Protein Science 3:1436
116. Eggen RIL, Geerling ACM, Waldkötter K, Antranikian G, de Vos WM (1993) Gene 132:143
116a. Kyrpides NC, Ouzonis CA (1995) Trends Biochem Sci 20:140
117. Bult CJ, White O, Olsen GJ, Zhou L, Fleischmann RD, Sutton GG, Blake JA, FitzGerald LM, Clayton RA, Gocayne JD, Kerlavage AR, Dougherty BA, Tomb JF, Adams MD, Reich CI, Overbeek R, Kirkness EF, Weinstock KG, Merrick JM, Glodek A, Scott JL, Geoghagen NSM, Weidman JF, Fuhrmann JL, Venter JC (1996) Science 273, 1058
117a. Sensen CW, Klenk HP, Singh RK, Allard G, Chan CC, Liu QY, Penny SL, Young F, Schenk ME, Gaasterland T, Doolittle WF, Ragan MA, Charlebois RL (1996) Mol Microbiol 22:175
118. Calvo JM, Matthews RG (1994) Microbiol Rev 58:466
119. Fitz-Gibbon S, Choi AJ, Miller JH, Stetter KO, Simon MI, Swanson R, Kim UJ (1997) Extremophiles 1, 36
119a. Pabo CO, Sauer RT (1992) Ann Rev Biochem 61:1053
120. Ciaramella M, Cannio R, Moracci M, Pisani FM, Rossi M (1995) World J Microbiol Biotechnol 11:71
121. Klenk HP, Palm P, Lottspeich F, Zillig W (1992) Proc Nat Acad Sci USA 89:407
121a. Ramos A, Raven N, Sharp R, Bartolucci S, Rossi M, Cannio R, Lebbink J, Van der Oost J, De Vos WM, Santos H (1997) 63:4020
122. Sutherland KJ, Danson MJ, Hough DW, Towner P (1991) FEBS Lett 282:132
123. Moracci M, La Volpe A, Pulitzer JF, Rossi M, Ciaramella M (1992) J Bacteriol 174:873
124. DiRuggiero J, Robb FT, Jagus R, Klump HH, Borges KM, Kessel M, Mai X, Adams MWW (1993) J Biol Chem 268:17767
125. Cannio R, de Pascale D, Rossi M, Bartolucci S (1994) Biotechnol Appl Biochem 19:233
126. O'Fagain C (1995) Biochim Biophys Acta 1252:1
127. Day MW, Hsu BT, Joshua-Tor L, Park J-B, Zhou ZH, Adams MWW, Rees DC (1992) Protein Science 1:1494
128. Baumann H, Knapp S, Lundback T, Ladenstein R, Hard T (1994) Nat Struct Biol 1:808
129. John J, Crennel SJ, Hough DW, Danson MJ, Taylor GL (1994) Structure 2:385
130. Russell RJ, Hough DW, Danson MJ, Taylor GL (1994) Structure 2:1157

131. Chan MK, Mukund S, Kletzin A, Adams MWW, Rees DC (1995) Science 267:1463
132. Hennig M, Darimont B, Sterner R, Kirschner K, Jansonius JN (1995) Structure 3:1295
132a. Aguilar CF, Sanderson I, Moracci M, Ciaramella M, Nucci R, Rossi M, Pearl L (1997) J Mol Biol 271:789
133. Yip KSP, Stilloman, Britton KL, Artymiuk PJ, Baker PJ, Sedelnikova SE, Engel PC, Pasquo A, Chiaraluce R, Consalvi V, Scandurra R, Rice DW (1995) Structure 3:1147
133a. Knapp S, De Vos WM, Rice D, Ladenstein R (1997) J Mol Biol 267:916
133b. Voorhorst WGB, Eggen RIL, Geerling ACM, Platteeuw C, Siezen RJ, De Vos WM (1997)
134. Doolittle WF, Lam LW, Schalkwyk LC, Charlebois RL, Cline SW, Cohen A (1992) Progress in developing the genetics of the *halobacteria*. In: Danson MJ, Hough DW, Lunt GG (eds) The Archaebacteria: biochemistry and biotechnology. The Biochemical Society, London, p 73
135. Koniski J (1989) Trends Biotechnol 7:88
136. Cammarano P, Teichner A, Londei P, Acca M, Nicolaus B, Sanz I, Amils R (1985) EMBO J 4:811
137. Grogan DW (1989) J Bacteriol 171:6710
138. Grogan DW (1991) J Bacteriol 173:7725
139. Kondo S, Yamagishi A, Oshima T (1991) J Bacteriol 173:7698
140. Aagaard C, Phan H, Trevisanato S, Garrett RA (1994) J Bacteriol 176:7744
141. Aagaard C, Leviev I, Aravalli RN, Forterre P, Prieur D, Garrett RA (1996) FEMS Microbiol Lett 18:89
142. Cubellis MV, Rozzo C, Montecucchi P, Rossi M (1990) Gene 94:89
143. Schleper C, Kubo K, Zillig W (1992) Proc Natl Acad Sci USA 89:7645
144. Charbonnier F, Erauso G, Barbeyron T, Prieur D, Forterre P (1992) J Bacteriol 174:6103
145. Elferink MGL, Schleper C, Zillig W (1996) FEMS Microbiol Lett 137:31
145a. Cannio R, Contursi P, Rossi M, Bartolucci S (1996) In: Thermophiles '96 Conference Abstracts. University of Georgia, USA, p 244
145b. Aravalli RN, Garrett RA (1996) Extremophiles 1:183
146. Steitz T (1990) Q Rev Biophys 23:229
147. Voorhorst WGB, Eggen RIL, Geerlink ACM, Platteeuw C, Siezen RJ, De Vos WM (1996) J Biol Chem 271:20426
148. Uemori T, Ishino Y, Toh H, Asada F, Kato I (1993) Nucl Acids Res 21:259
149. Halio SB, Blumentals II, Short SA, Merrill BM, Kelly RM (1996) J Bacteriol 178:2605
150. Robinson KA, Schreier HJ (1994) Gene 151:173
151. Pedroni P, Della Volpe A, Galli G, Mura GM, Pratesi C, Grandi G (1995) Microbiology 141:449
152. Hethke C, Geerling ACM, Hausner W, De Vos WM, Thomm M (1996) Nucl Acids Res 12:2369
153. Schut G, Kengen SWM, Hagen WR, Dahlke I, Thomm M, Van der Oost J, De Vos WM (1997) (submitted)
154. Douglas R et al. (1997) J Bacteriol 179:7135
155. Klenk HP et al. (1997) Nature 390:364

Received August 1997

An Overview of the Role and Diversity of Compatible Solutes in *Bacteria* and *Archaea*

M. S. da Costa[1] · H. Santos[2] · E. A. Galinski[3]

[1] Departamento de Bioquímica, Universidade de Coimbra, 3000 Coimbra, Portugal
E-mail: milton@cygnus.ci.uc.pt
[2] Instituto de Tecnologia Química e Biológica, Universidade Nova de Lisboa, Rua da Quinta Grande 6, Apartado 127, 2780 Oeiras, Portugal
[3] Institut für Mikrobiologie & Biotechnologie, Rheinische Friedrich-Wilhelms-Universität Bonn, Meckenheimer Allee 168, 53115 Bonn, Germany

The accumulation of compatible solutes is a prerequisite for the adaptation of micro-organisms to osmotic stress imposed by salt or organic solutes. Two types of strategies exist to cope with high external solute concentrations; one strategy is found in the extremely halophilic *Archaea* of the family *Halobacteriaceae* and the *Bacteria* of the order *Haloanaerobiales* involving the accumulation of inorganic ions. The other strategy of osmoadaptation involves the accumulation of specific organic solutes and is found in the vast majority of microorganisms. The organic osmolytes range from sugars, polyols, amino acids and their respective derivatives, ectoines and betaines. The diversity of these organic solutes has increased in the past few years as more organisms, especially thermophilic and hyperthermophilic *Bacteria* and *Archaea*, have been examined. The term compatible solute can also be applied to solutes that protect macromolecules and cells against stresses such as high temperature, desiccation and freezing. The mechanisms by which compatible solutes protect enzymes, cell components and cells are still a long way from being thoroughly elucidated, but there is a growing interest in the utilization of these solutes to protect macromolecules and cells from heating, freezing and desiccation.

Keywords: Compatible solutes, osmolytes, salt stress, temperature stress, *Bacteria*, *Archaea*.

Advances in Biochemical Engineering/
Biotechnology, Vol. 61
Managing Editor: Th. Scheper
© Springer-Verlag Berlin Heidelberg 1998

1
Introduction

1.1
What Is an Extremophile?

The evolution, physiology, biochemistry, and the biotechnological applications of the so-called extremophilic microorganisms are very popular and important subjects of current scientific investigation. The organisms that grow at extremely high or low temperatures, at extreme pH values, under conditions of osmotic stress, that are resistant to high gamma radiation, grow under high hydrostatic pressures among other conditions, are generally considered to be extremophilic. Extremophiles can also be defined as organisms that thrive in environments where biodiversity is apparently low due to conditions that inhibit the growth of many other species. The apparent decrease in diversity as the environment becomes progressively more extreme could be due to our lack of knowledge of the number of different species that inhabit these environments. This is no doubt true and new extremophilic species are continually being described, but it is unlikely that the number of species that inhabit extreme biotopes will ever be as large as those in environments that are more common in the biosphere.

In this review we will discuss the strategies that enable microorganisms to cope with fluctuations in the concentration of small molecular mass solutes in the environment, and particular attention will be given to the variety of intracellular organic solutes that accumulate in these organisms in response to an increase in the external solute concentration. We will not, however, restrict the discussion to extremophilic organisms because so much of the knowledge on osmotic adaptation was obtained from microorganisms that are not considered extremophilic. We will also glance at results that show some organic solutes to have a role in protecting the cell and cellular components from other types of stresses such as high temperatures, freezing and desiccation.

1.2
Overview of Terms and Concepts

Microorganisms can be found growing in aqueous environments ranging from fresh water to saturated brines or extremely concentrated sugar solutions. Nevertheless, the vast majority of the microbial species described originate from aqueous environments which are dilute in terms of salts or sugars. Yet most microorganisms examined appear to be able to adjust physiologically, within limits of intrinsic tolerance, to variations in the levels of osmotically active substances that occur in aqueous environments and that influence growth or survival. On the other hand, many species are, in fact, adapted to thrive under conditions of osmotic stress and require NaCl or high concentrations of sugars for growth.

An increase in the concentration of low molecular mass solutes of an aqueous environment always results in a decrease in the water available to the microorganism. The decrease in the external water leads to a decrease in the cell volume and/or the turgor pressure ultimately affecting metabolic systems and macromolecules [1]. In order to adjust to the higher solute concentrations of the environment, microorganisms must accumulate an intracellular solute to re-establish the cell turgor pressure and/or cell volume and protect the activity of intracellular enzymes and other macromolecules [1, 2]. The failure to adjust osmotically to higher solute concentrations in the environment will, of course, result in the cessation of growth or death.

Many authors have used water activity (a_w) to describe the amount of water that is thermodynamically available to the cell, expressed as its mole fraction, because of the conceptual simplicity of this expression, the relative ease of experimental measurement of water activity or extrapolation of values for a few solutes from established tables, and its independence from temperature [1, 3, 4].

Under conditions of water stress which depends, of course, on the particular organism and the solute used to cause the stress, the accumulation of an intracellular solute can reach levels that interfere with the organism's metabolism; therefore, the accumulation of an osmolyte must be compatible with cellular metabolism and must affect the properties of cellular components as little as possible. Moreover, as will be shown later, the concentration of an osmolyte can reach enormous levels. To convey the idea that osmolytes cannot interfere with cell function Brown coined the term compatible solute [5]. This term is applicable to intracellular solutes that preserve metabolism, and protect cells and cell components under conditions of water stress, and will be used throughout this review even though the term osmolyte is applicable in most cases and will also be used.

The term compatible solute was initially applied to organic and inorganic osmolytes [5]. However, this term is generally applied to organic osmolytes that protect non-salt dependent macromolecules found in the majority of organisms from the inhibitory effect of inorganic ions or many organic molecules. It can, nevertheless, be argued that the accumulation of potassium, for example, in the organisms whose macromolecules are salt-dependent is better suited for stability and activity than are other inorganic ions such as sodium. In this case, potassium could also be considered a compatible solute.

Furthermore, as we will see later, the term compatible solute can be extended to solutes that protect cells from stresses that appear to be unrelated to water stress but may be, in many cases, related to alterations in the amount and/or the properties of intracellular water. Some organic solutes may be able to protect macromolecules and cells from freezing, desiccation, and high temperature. The term compatible solute will, therefore, also be applied to solutes that protect macromolecules and cells from these stress conditions.

The variety of small molecular weight solutes that can serve as compatible solutes is constrained by the ability of the solute to protect cell function [6]. The number of compatible solutes is relatively small, and they fall into certain categories of compounds such as polyols and derivatives, sugars and derivatives, amino acids and amino acid derivatives, betaines, and ectoines [2]. Most compatible solutes are widespread in microorganisms capable of osmotic adjustment but there are, of course, compatible solutes that are restricted to one known organism or a small number of organisms. The number of different compatible solutes has, nevertheless, increased in the past few years as more organisms are studied, and new solutes that serve as osmoprotectants or thermoprotectants are likely to be discovered as thermophilic and hyperthermophilic organisms are increasingly studied.

The term osmoregulation has been also generally used to describe the capacity of the cell to adjust physiologically to water stress, but the term osmoadaptation has been used recently because it has a broader meaning and conveys the immediate physiological and genetic alterations that take place in the cell as the level of environmental water changes [2, 7].

Osmoadaptation in *Bacteria* and *Archaea* has been generally examined in media where the water activity is decreased by the addition of salt, so that the water relations of these microorganisms are classified in terms of their relationship to NaCl in the environment. The term halotolerant is used for organisms that do not require more than minor levels of Na^+ for growth, but which can grow in environments with salt concentrations that inhibit the growth of less tolerant organisms. Organisms that require Na^+ for growth are classified as slightly halophilic, moderately halophilic, or extremely halophilic depending on the range of NaCl which allows growth [8, 9]. This review will be primarily devoted to osmoadaptation of *Bacteria* and *Archaea* to water stress imposed by salt, but we must not forget that many eukaryotic organisms and a few bacteria are also well adapted to cope with water stress imposed by sugars. Moreover, yeasts and filamentous fungi are unsurpassed in their ability to cope with low water activity imposed by sugars [1, 10]. To describe the water relations of yeast and fungi which can be salt- and sugar-tolerant, terms using the prefixes osmo- or xero- have been coined and are used in the literature [1, 5]. For example, osmotolerant or xerotolerant yeasts do not require large amounts of salt or sugars in the medium, but can grow in media with elevated levels of salts or sugars, while osmophilic or xerophilic organisms require NaCl or sugars in high concentrations to grow.

1.3
Strategies for Osmoadaptation

There are two principal mechanisms of osmoadaptation in bacteria, archaea and eukaryotes that demonstrate two completely different evolutionary strategies that have been adopted to cope with the same type of stress. One strategy for maintaining osmotic equilibrium across the membrane involves the influx of salts into the cytoplasm and has been called the salt-in-the-cytoplasm or the saline cytoplasm type of osmoadaptation. [2, 11]. These organisms can accumulate enormous quantities of inorganic ions (K^+, Na^+, Cl^-) and have, during the course of evolution, developed proteins, and other macromolecules that cope with, and take advantage of, the high intracellular salt concentrations. A highly saline cytoplasm is found in two completely unrelated groups of organisms, the extremely halophilic *Archaea* of the family *Halobacteriaceae* and the anaerobic halophilic *Bacteria* of the order *Haloanaerobiales* [1, 12–15]. This form of osmoadaptation co-evolved with the structural modification of a large number of cellular components. The composition of enzymes is the most notable effect of the intracellular saline environment on macromolecular structure, but other components such as ribosomes are equally modified. Most enzymes studied have a negative charge due to the predominance of acidic amino acids over basic amino acids, which has been explained by the necessity to attach a strong hydration shell around proteins. Moreover, most of the enzymes from extremely halophilic archaea have a complete dependence on K^+ and/or Na^+ for activity.

The other type of osmoadaptive strategy found in the vast majority of the microorganisms examined involves the accumulation of specific organic osmolytes with the exclusion of salt and the majority of the organic solutes found in the environment. In these organisms intracellular macromolecules have not undergone specific modifications and are, therefore, sensitive to high intracellular concentrations of salts and most organic solutes. However, this mechanism, greatly reduces the necessity to modify genetically the enzymatic and structural makeup of the cell, providing a very versatile means for rapid adaptation to an osmotically fluctuating environment [2]. Perhaps, for this reason, the accumulation of organic osmolytes is widespread in nature.

The extremely halophilic *Archaea* are viewed as the archetypal halophiles, and it is sometimes supposed that they are unsurpassed in their ability to cope with salt. However, this mechanism of osmoadaptation does not appear to be advantageous over the accumulation of organic osmolytes, since some species, such as the microalgae of the genus *Dunaliella* that accumulate glycerol [16, 17], have been isolated from environments similar to those inhabited by extremely halophilic *Archaea* and can also grow in saturated NaCl media [1, 16]. Moreover, the known species of the *Haloanaerobiales* are less halophilic than many microorganisms that accumulate organic osmolytes.

2
Compatible Solutes

2.1
Halophilic Eubacteria

2.1.1
Potassium and Glutamates

Early reports on low level osmoresponse, using *E. coli* and *Salmonella typhimurium* as model systems, were able to demonstrate a linear increase in cytoplasmic potassium concentration with increasing salinity of the medium [18]. It was, therefore, assumed that potassium accumulation was largely responsible for osmotic balance across the membrane and accounted for osmotic equilibrium in many non-halophilic Proteobacteria [19]. The resulting excess positive charge is, however, not compensated by chloride, as in the *Halobacteriaceae* and the *Haloanaerobiales*, but is almost completely balanced by organic anions, either synthesized or accumulated from the medium [20]. The most important charge counterbalance for raised cytoplasmic potassium levels is provided by glutamate isomers (Fig. 1), in particular α-glutamate [21]. Its osmotic regulation has been well documented and a substantial increase from approximately 30 to 300 mmol/l (when the external osmolality is raised to that of sea water) appears to be a general phenomenon, at least for Gram-negative bacteria [22]. The upper limit for cytoplasmic glutamate levels seems to be around 500 mmol/l as calculated from the total cytoplasmic water [23, 24], and is often reached only transiently before the onset of subsequent adaptational mechanisms. Because of its central role in bacterial metabolism the precise regulation of glutamate levels is probably difficult to resolve, although there has been a suggestion that pH or potassium triggers its synthesis via the activation of glutamate dehydrogenase [22]. The zwitterionic decarboxylation product of α-glutamate, γ-aminobutyric acid (Fig. 1), has been reported in some cases, but its involvement in osmoadaptation is, at present, still unclear. The isomeric β-glutamate (Fig. 1), a relatively rare amino acid, has been observed in marine bacteria [25], a number of methanogenic bacteria such as thermophilic *Methanococcus* species, mesophilic *Methanogenium* species [26], all *Nocardiopsis* species [11], and in *Thermotoga neapolitana* and *T. maritima* [27]. The dependence of β-glutamate levels on salt concentration in the medium strongly suggests a function as an osmolyte, but again a cytoplasmic concentration limit which is similar in range to that of α-glutamate seems to preclude a function as a "real" compatible solute. Another glutamate derivative, α-glutamate betaine (Fig. 1), has thus far only been reported in the cyanobacterium *Calothrix* N181 [28]. It may serve as an example of another type of betaine, but in contrast to glycine betaine and dimethylsulfoniopropiothetine (DMSP) this compound is negatively charged at physiological pH.

Potassium in combination with organic anions (e.g. glutamates) is apparently only suitable for a low level osmoresponse to salinities not much higher than those of sea water. At salinities well above sea water additional organic osmo-

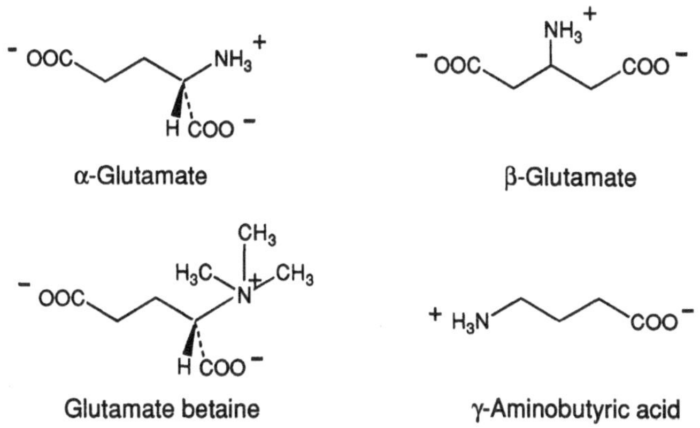

Fig. 1. Glutamate and derivatives

lytes are needed to achieve osmotic equilibrium and to stabilize the cytoplasm in a low water environment.

Results for Gram-positive bacteria suggest higher cytoplasmic potassium levels, which are largely unaffected by changes in osmolarity. Therefore, a primarily different role of potassium on turgor regulation cannot be excluded here [29–32].

2.1.2
Sugars

Sucrose and/or trehalose are widespread disaccharides occurring in eukaryotic and prokaryotic organisms alike (Fig. 2). Because they are often observed in slightly halotolerant or marine organisms growing at 3% NaCl (0.5 mol/l) or just above, a prospective role as osmolytes was assumed. Examples of organisms accumulating sucrose are non-halophilic cyanobacteria, while the use of trehalose appears to be common in many bacterial groups, including a wide range of cyanobacteria [28, 33], anoxygenic phototrophic species [34], as well as other representative bacteria such as *Azotobacter chroococcum, Klebsiella pneumoniae* [35], *Pseudomonas aeruginosa* [36], *Rhizobium meliloti* [37] and *E. coli* [38]. The latter has subsequently become a model organism for in-depth investigation into the osmoregulatory response mechanism [23, 24, 39, 40], and a sequence of events has been elucidated involving potassium, α-glutamate and trehalose. The proposed role of trehalose as a more compatible replacement for the rapidly accumulated potassium and synthesized glutamate was subsequently challenged by others who observed a strong influence of growth conditions, and growth phase on trehalose accumulation. Only under nitrogen limitation did trehalose appear to replace glutamate, but not potassium. Although trehalose, without doubt, participates in some way to alleviate osmotic stress, there seems to be increasing "consensus" that its involvement in osmoadaptation is only minor, and that trehalose is primarily a stress metabolite associated with the onset of unfavorable growth conditions and designed to ensure survival [41, 42]. As

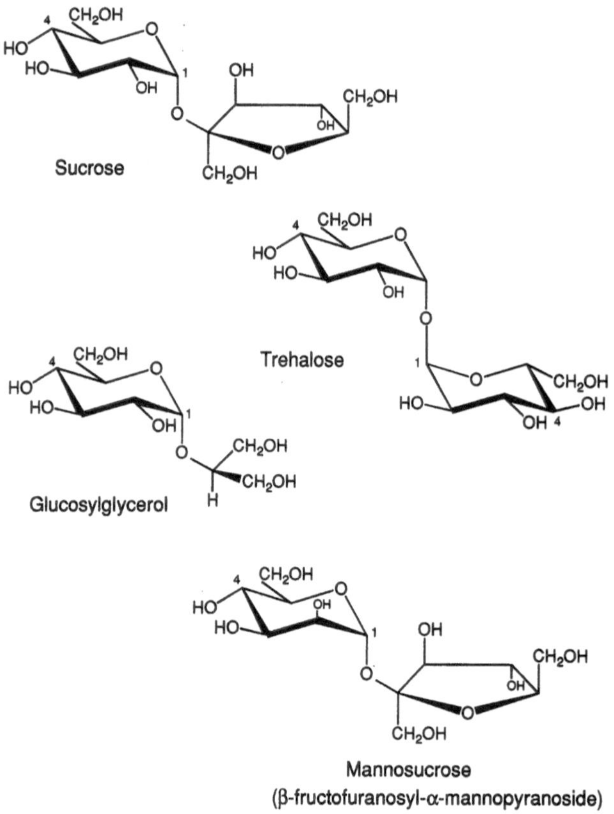

Fig. 2. Sugars and heterosides

shown by molecular biology studies in *E. coli*, trehalose synthesis is, in fact, under the control of *rpo*[s], a gene encoding sigma factor σ^s, which is responsible for the expression of a large number of genes induced upon entry of stationary phase [43].

High amounts of trehalose are often found in resting stages such as spores or other anhydrobiotic forms of life. Trehalose has been termed the carbohydrate of dormancy because it supplies energy in the resting stage and confers tolerance towards freezing and drying [44–46]. This non-reducing disaccharide is known to stabilize membranes, proteins and whole cells during dehydration and storage [47–49], possibly due to water replacement mechanisms at the membrane level, its excellent "glass forming" abilities and a potential role as free radical scavenger (as trehalose derivatives) during anabiosis. Considering that dehydration indeed often coincides with increasing salinity, possibly anticipated by an osmo/salt sensor, it does not seem surprising that trehalose is often found in really halophilic bacteria as a minor component (<500 mmol/l) in combination with other more "potent" compatible solutes such as betaine or ectoine [50].

An unusual disaccharide mannosucrose (*β*-fructofuranosyl-*α*-mannopyra-noside) (Fig. 2) has been detected in *Agrobacterium tumefaciens* biotype I which is tolerant up to 2% NaCl, where it seems to take over a similar role as a stress protectant. The choice of this rare sugar is explained in terms of the ecology of the organism, which must avoid both sucrose and trehalose, the former because it functions as an attractant towards the rhizosphere and the latter because it is toxic to many angiospermic plants [51].

2.1.3
Compatible Solutes "Sensu Stricto"

A fairly comprehensive survey of phototrophic and chemoheterotrophic eubacteria covering the eubacterial divisions has revealed a broad spectrum of compatible solutes [50, 52–65]. These osmolytes usually reach cytoplasmic concentrations above 500 mmol/l, fulfill a number of prerequisites, namely high solubility and charge neutrality at physiological pH, and fall into the following classes of compounds:

- polyols and heterosides;
- amino acids (proline, alanine, glutamine and derivatives);
- N-acetylated diamino acids ($N\delta$-acetyl–ornithine, $N\varepsilon$-acetyl–lysine)
- ectoines (ectoine, β-hydroxyectoine);
- betaines (trimethylammonium compounds) and thetines (dimethylsulfonium compounds).

2.1.3.1
Polyols and Heterosides

Thus far the de novo biosynthesis of polyols like glycerol, arabitol, and inositol, which are typical compatible solutes of halophilic/halotolerant fungi, algae and plants [17, 66, 67], has nearly been observed in bacteria, although two species employ polyols for osmoadaptation.

The natural environment of *Zymomonas mobilis* includes sugary fruit saps and honey, while being very sensitive to NaCl. During growth on high levels of sucrose *Z. mobilis*, unlike all other heterotrophic bacteria known, converts sucrose to glucose and sorbitol, the latter of which serves as a compatible solute. Other compatible solutes commonly found in bacteria, such as betaine, are not found in cells derived from a medium containing yeast extract where this osmolyte is present. Sorbitol added in low concentrations (50 mmol/l) to high–glucose medium (exceeding 0.83 mol/l) is also taken up improving growth significantly at low water activity imposed by glucose. The accumulation of a polyol by *Z. mobilis* may represent an interesting example of convergent evolution in osmoadaptation, because this species, like many yeasts, has developed a means of modifying sugars present in high levels in the environment to produce a compatible solute [68]. On the other hand, *Pseudomonas putida* was found to accumulate mannitol (and $N\alpha$-acetyl–glutaminylglutamine-1-amide) under osmotic stress [69], indicating that the accumulation of polyols by some bacteria may be more common than previously suspected.

The only polyol derivative detected in bacteria is O-α-D-glucopyranosyl-$(1\rightarrow2)$-glycerol (Fig. 2), a heteroside structurally related to floridosides (O-α-Dgalactopyranosyl-$(1 \rightarrow 2)$-glycerol) and isofloridosides (O-α-D-galactopyrano-syl-$(1 \rightarrow 1)$-glycerol) of certain red algae and chrysophyceae [70, 71]. Glucosyl-glycerol has been known for a long time as the typical osmolyte of a wide range of moderately halophilic cyanobacteria with intermediate salt tolerance [28, 33, 52, 53, 72]. In contrast to the biosynthetic pathway of heterosides in red algae (condensation of UDP-sugar with glycerol phosphate) [73], cyanobacterial glucosylglycerol is in fact synthesized from ADP-activated glucose and glycerol-3-phosphate in *Synechocystis* sp. PCC6803 [74–76]. Glucosylglycerol was sub-sequently also detected in the moderately halophilic purple bacterium *Rhodobacter sulfidophilus* [11, 50, 56] and in the slightly halotolerant species *Pseudomonas mendocina* and *P. pseudoalkaligenes* [77]. It appears, therefore, that heterosides probably constitute a class of widespread compatible solutes, typically connected with intermediate salt tolerance.

2.1.3.2
Natural Amino Acids

Apart from glutamate and suitable derivatives, the natural amino acids alanine, glutamine and proline are important bacterial osmolytes (Fig. 3). While the first two are definitely low level adjusters (particularly for Gram-positive bacteria) proline can accumulate to very high concentrations. It is also often observed that all three components are raised in a salt stress situation, as in *Streptomyces* sp. [78]. α-Glutamine has a relatively low solubility (approx. 0.3 mol/l) compared to other solutes (proline: 14 mol/kg) and, therefore, high cytoplasmic concentrations reported in moderately halotolerant members of the genus *Corynebacterium* [64] come close to saturation. The β-form of glutamine, which has so far only been detected in halophilic methanogens, reaches cytoplasmic concentrations well above 0.5 mol/l in some species [26, 79–81], and it is probably this higher solubility (but not the solubility alone) which makes this compound a suitable compatible solute. Similar strategies seem to apply for α-glutamine derivatives containing glutamine-1-amide as the characteristic structural element. This unique class of compounds (polar, but not zwitterionic), is typically carbamoy-lated or acetylated at the α-nitrogen, or in the case of dipeptides, linked to the carboxyl terminus of a second amino acid (Fig. 3). While $N\alpha$-carbamoylgluta-mine-1-amide (Fig. 3) has so far only been described as a minor component in *Ectothiorhodospira marismortui* [82], the neutral dipeptide $N\alpha$-acetyl–gluta-minylglutamine-1-amide (Fig. 3) is typical of a number of moderately halophil-ic purple bacteria [84] and other Proteobacteria such as *Azospirillum brasilense* [11], *Rhizobium meliloti* [37] and *Pseudomonas putida* [69]. With a maximum cytoplasmic concentration well below 1.0 mol/l the glutamine-1-amides take an intermediary position between typical low level osmoadaptors and real compa-tible solutes.

None of the above natural amino acids (and derivatives) can rival the impor-tance of proline (Fig. 3) as the outstanding natural amino acid-type osmolyte, which can be accumulated to molar concentrations and comprise as much as 20% of the cells' dry weight. Within prokaryotes, proline was originally regarded as the typical osmolyte of halophilic *Bacillus* species, a view primarily based on

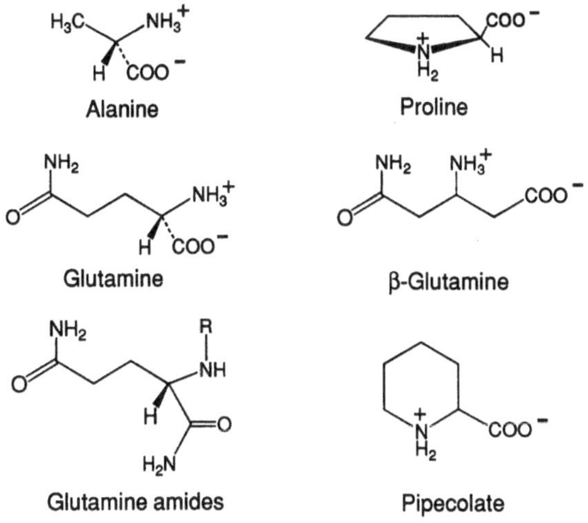

Alanine

Proline

Glutamine

β-Glutamine

Glutamine amides

Pipecolate

Fig. 3. Natural amino acids and derivatives. R=-CONH$_2$ (*N*α-carbamoylglutamine-1-amide) or -*N*α-acetylglutaminyl (*N*α-acetyl-glutaminylglutamine-1-amide)

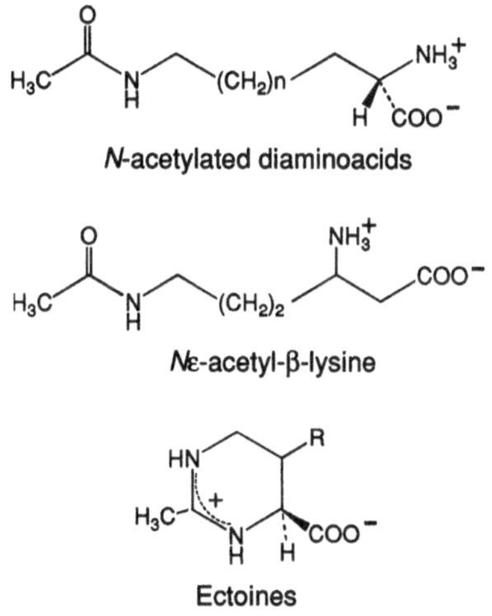

N-acetylated diaminoacids

*N*ε-acetyl-β-lysine

Ectoines

Fig. 4. *N*-acetylated diamino acids and ectoines. *N*δ–acetyl-ornithine (n=1), *N*ε-acetyl-lysine (n=2), L-ectoine (R=H), β-hydroxyectoine (R=OH)

investigations into *B. subtilis* and closely related species. It was only shown later that the majority of halophilic/halotolerant bacillus-type organisms produce ectoine, either alone or in combination with proline and/or acetylated diamino acids [50, 61]. *Bacillus subtilis* and *Planococcus citreus*, therefore, seem to belong to a minority of proline osmoadapted bacillus-type organisms, which are unable to synthesize other osmolytes. Further proline producers are found among *Staphylococcus* and *Salinicoccus* species. The fact that these organisms are typically moderately halophilic and truly halotolerant appears to support the view that proline is a solute of lesser compatibility, at least at extreme salt concentrations. In this context, it is also noteworthy that the compatible solute pool of proline producers may vary remarkably with media composition and growth conditions. Strain M96/12b, for example, upon entering the stationary phase converts proline into another compatible solute, $N\delta$-acetyl-ornithine (Fig. 4) [60]. The proline homologue, pipecolic acid (Fig. 3), has been described as an osmolyte in *Corynebacterium ammoniagenes* (renamed *Corynebacterium glutamicum*) [62]. This observation, however, was later questioned [99] and it remains to be seen if pipecolic acid is synthesized de novo or preferentially accumulated from the medium.

2.1.3.3
N-Acetylated Diamino Acids and Ectoines

N-Acetylation of diamino acids such as ornithine and lysine converts a positively charged amino acids into a neutral zwitterionic solute and thus into a more compatible osmolyte. The role of $N\delta$-acetyl-ornithine in osmoadaptation was first documented with strain M96/12b [60], which is related to typical bacillus-type organisms. Subsequently, this compatible solute was also detected, at least, in minor amounts, in almost all *Bacillus* species under investigation, and also *Sporosarcina halophila* and *Planococcus citreus* [50]. Similarly, the homologous $N\varepsilon$-acetyl-lysine (Fig. 4), originally isolated and identified from *Sporosarcina halophila*, was also shown to be relatively widespread among bacilli and related organisms [50, 61]. Shorter homologues than acetylated ornithine, namely N-acetylated diaminobutyric acid, have thus far not been detected in halophilic bacteria, except as precursors during ectoine biosynthesis.

Ectoine (Fig. 4) was named after its discovery in the phototrophic sulfur bacterium *Ectothiorhodospira halochloris* [55]. According to its chemical structure, ectoine can be classified as a partially hydrated pyrimidine carboxylic acid (1,4,5,6-tetrahydro-2-methyl-4-pyrimidine carboxylic acid). Its pathway of synthesis, which was investigated in *Ectothiorhodospira halochloris* and *Halomonas elongata* [56, 83–85] revealed a biosynthetic sequence proceeding via aspartic semialdehyde, diaminobutyric acid and $N\gamma$-acetylated diaminobutyric acid (Fig. 5). These enzymatic steps require only three additional enzymes: diaminobutyric acid transaminase, diaminobutyric acid acetylase and $N\gamma$-acetyl-diaminobutyric acid dehydratase (ectoine synthase). Ectoines may, therefore, be seen as cyclic forms of N-acetylated diamino acids. Because of the delocalization of the π-electrons the amino functions are not reactive and, therefore, not detectable by standard amino acid analysis, which explains why these compounds escaped detection for such a long time. Ectoines are, without doubt, one of the most abundant osmolytes in nature, and as much as glycine betaine can be regarded the typical product of halophilic phototrophic bacteria. Ectoines can be seen as the common solutes of aerobic heterotrophic eubacteria, and are found in all truly halophilic Proteobacteria, *Nocardiopsis* and *Streptomyces* species, brevibacteria, micrococci, bacillus-like organisms (except *B. subtilis* and *Planococcus citreus*) and *Marinococcus* sp. [11, 50, 64]. Although ectoines were first discovered in halophilic bacteria, where their osmotic

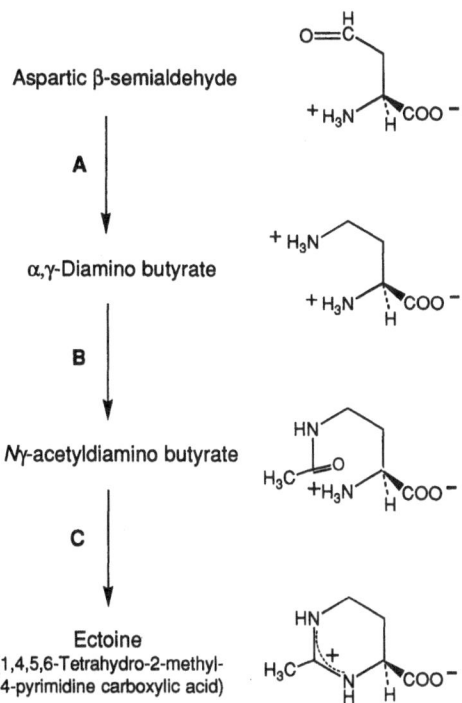

Fig. 5. Biosynthesis of ectoine. 2,4-diaminobutyrate transaminase (A), 2,4-diaminobutyrate acetylase (B), and Nγ-acetyldiaminobutyrate dehydratase (ectoine synthase) (C)

function is incontestable, they have subsequently been found in non-halophilic antibiotic-producing streptomycetes, where they seem to protect the producer strain from its own antimicrobial product [86]. This suggests that ectoines or deduced derivatives may serve a variety of functions in nature. As ectoines are, at present, not available through chemical synthesis, and their biotechnological production from halophiles has become a major focus of interest [87, 88].

2.1.3.4
Betaines and Thetines

Glycine betaine (subsequently named betaine) appears to be a truly universal organic osmolyte (Fig. 6) occurring in salt and sugar tolerant higher plants (e.g. *Beta vulgaris*), algae, the animal kingdom [89–95] and prokaryotes of different ecophysiology. Its biosynthetic pathway had already been investigated in halophytes of the family Chenopodiaceae [96] when betaine was discovered as a compatible solute in bacteria [54]. An analogous biosynthetic route via the oxidation of choline was, however, ruled out for *Ectothiorhodospira* species [56, 58]. Instead, a reaction pathway via the methylation of glycine, which also uses S-adenosyl methionine as a methyl donor, was elucidated for *Ectothiorhodospira halochloris* [97] and phylogenetically distant archaeal *Methanohalophilus* species [81, 98].

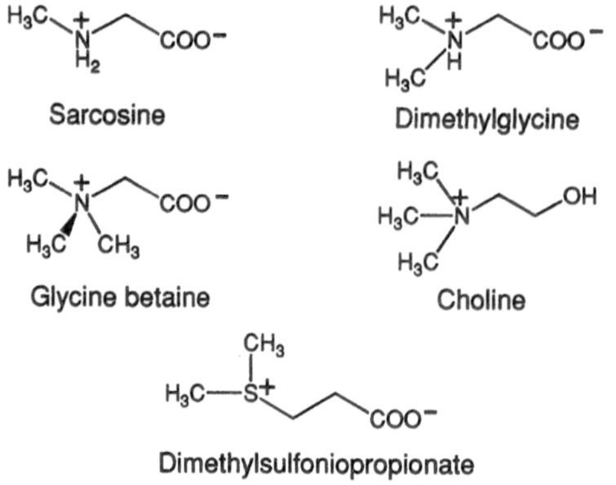

Fig. 6. Betaines and thetines

Although the role of betaine as a universal compatible solute is undoubted, the ability to synthesize betaine was largely overestimated in the past [99]. This is explained by the fact that betaine (or suitable precursors) are common constituents of many complex media components (for example, yeast extract), and that the organisms' transport capacities for compatible solutes, and betaine in particular, were not known at the time. In addition, there has been confusion in the literature, where the term "betaine synthesis" was used indiscriminately to describe a two-step oxidation of the precursor choline involving a conversion process rather than de novo biosynthesis. Today we understand that the ability to synthesize betaine de novo is rare among chemoheterotrophic eubacteria, one of the few exceptions being *Actinopolyspora halophila*, but is common in both oxygenic and anoxygenic phototrophic eubacteria, and methanogens, displaying moderate or high salt tolerance [11, 28, 53, 54, 100].

The methylated sulfur analogues of betaines are the thetines (Fig. 6), in particular dimethylsulfoniopropiothetine (DMSP). This compound is common in the marine environment, and a role as an osmotic solute has been proposed for a number of plants, cyanobacteria and algae [93, 94, 101]. Little is known about the biosynthetic pathway of DMSP in microorganisms, except for the fact that methionine is the source of the sulfur atom and of both methyl groups in the green alga *Ulva lactuca*, and that the α-carbon of methionine yields the carboxyl group of DMSP [94, 101]. These data led to the proposal of a pathway where methionine is oxidatively deaminated, decarboxylated and finally methylated [102]. However, the actual sequence of reactions has so far not been clarified in microorganisms.

2.2
Methanogens (Amino Acid-Type Solutes and Potassium)

Interest in halophilic methanogens began with the isolations of strains from saline environments, which used trimethylamine as a substrate for methanogenesis [103–106]. The occurrence of betaine in many salt stressed methanogens was, at first, observed as the result of accumulation from yeast extract-containing media [79], but finally de novo synthesis was also reported in many halophilic *Methanohalophilus* strains. These organisms accumulated betaine to a concentration of 0.7 mol/l when grown on trimethylamine, and of 1.7 mol/l when the organism is grown on methanol in medium containing about 16% NaCl [80, 98]. Betaine biosynthesis was shown to proceed via the glycine methylation pathway (SAM) and, as such, probably presents a striking example of convergent evolution of osmolyte systems [81]. Some *Methanohalophilus* strains, which apparently did not synthesize betaine, were shown to produce the precursor dimethylglycine instead [107]. It is presently not known if an ineffective final methylation step is responsible for the accumulation of dimethylglycine. In any case, the dimethyl variant of betaine is apparently equally suitable for osmotic adaptation.

In addition to β-glutamate, and β-glutamine Nε-acetyl-β-lysine (Fig. 4) has been described as a compatible solute in halophilic *Methanococcus* and *Methanohalophilus* strains [26, 80, 108]. Determination of turnover rates from pulse chase experiments has also shown that Nε-acetyl-β-lysine is metabolically relatively "inert" (t $_{0.5}$ = 12 h, as compared to 3 h for α-glutamate and 32 h for betaine), which is consistent with its role as an osmolyte in methanogens [137]. The labeling pattern for this compound suggests a synthesis of the α-precursor occurring via the diaminopimelate pathway [81].

With respect to the role of potassium in osmoadaptation of methanogens the following types of strategies seem to emerge. Non–marine *Methanobacterium* spp. such as *Methanobacterium thermoautotrophicum* strains ΔH and Marburg show no evidence of solute accumulation in response to increasing osmotic pressure, but generally display a hypertonic cytoplasm (approx. 800 mmol/l K$^+$) which enables them to grow up to a salinity of 0.5 mol/l NaCl [109, 110]. As a counterbalance for potassium the anionic compound cyclic 2,3-bisphosphoglycerate (cBPG) and 1,3,4,6 hexanetetracarboxylic acid (HTCA) have been described as important components in the low molecular mass solute pool in *Methanobacterium thermoautotrophicum* ΔH and Marburg (Fig. 7) [111]. In particular HTCA (in contrast to glutamate) was shown to have a slow metabolic turnover, which led the authors to propose a function as an osmolyte for this solute. Under optimal growth conditions (60–65°C, 10 mmol/l external NaCl) the cytoplasmic concentration of HTCA was rather small (32 mmol/l) compared to glutamate (96 mmol/l) and cBPG (172 mmol/l). Nevertheless, it contributed significantly towards the pool of counterions balancing the high cytoplasmic K$^+$ levels.

Moderately halophilic and marine *Methanosarcina* species, which display a salt tolerance of up to 1 mol/l NaCl, accumulate potassium and glutamate to a threshold level of 0.5 mol/l, which is similar to the response reported for enterobacteria. When the external osmolality approaches 1.0 mol/l Nε-acetyl-β-lysine is synthesized and, subsequently replaced, by betaine accumulated from the medium, if present [112]. For the hyperhalophilic *Methanohalophilus* strain Z7302, on the other hand, an increase of internal potassium from 1 to 3 mol/l was demonstrated when the external salinity was raised from 2 to 4 mol/l NaCl with a concomitant increase in the betaine levels from 0.3 to 3.8 mol/l. These

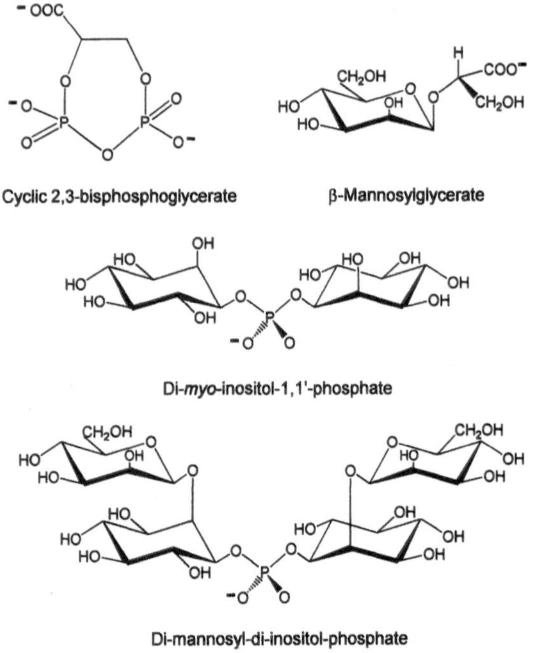

Cyclic 2,3-bisphosphoglycerate   β-Mannosylglycerate

Di-*myo*-inositol-1,1'-phosphate

Di-mannosyl-di-inositol-phosphate

Fig. 7. Novel organic solutes from methanogens, thermophiles and hyperthermophiles

observations led the authors to propose that both potassium ions and betaine are the major compatible solutes in these extremely halophilic methanogens [113]. Some methanogens, therefore, seem to employ a strategy of adaptation which uses elements of the compatible solute strategy and of the salt-in-the-cytoplasm type osmoadaptation.

Other organic solutes of methanogens also deserve mention here. Aspartate has been found to increase slightly over the range of 0.2 to 1.0 mol/l NaCl in *Methanococcus thermolithotrophicus*, although the primary compatible solutes were α- and β-glutamate. The slight increase in the intracellular levels of aspartate in response to salt stress shows that this amino acid contributes to the compatible solute pool of the organism, and aids in balancing the positive charge of potassium [114].

The heterosides, as previously discussed, are compatible solutes in a wide range of cyanobacteria and algae. Recently, a closely related compound was identified as α-glucosylglycerate in *Methanohalophilus* strain FDF1 [98]. Betaine, Nε-acetyl-β-lysine and α-glutamate are the primary osmolytes of this organism, while α-glucosylglycerate did not behave as a compatible solute and was believed to be a metabolic intermediate. However, as we will discuss later, a similar compound, mannosylglycerate, acts as a compatible solute in thermophilic bacteria and hyperthermophilic archaea, indicating that glucosylglycerate may serve a similar function under other growth conditions.

As mentioned above, cBPG serves as counterbalance for potassium but its function is, at present, controversial. This solute has only been detected in methanogens, such as *Methanobacterium thermoautotrophicum, Mb. bryantii, Methanobrevibacter smithii, Methanothermus fervidus* and *Mt. sociabilis, Methanosphaera stradtmanae*, and *Methanopyrus kandleri* [115–119]. The optimum growth temperature of these species varies from 37 °C for *Mb. bryantii* to 88 °C for *Mt. fervidus* and 98 °C for *Mp. kandleri*. In most of these organisms there seems to be a positive correlation between the levels of intracellular cBPG, K^+ and high growth temperatures culminating in the accumulation of very large concentrations of cBPG in *Mp. kandleri* [119]. In some strains, the K^+ levels are higher than those of cBPG and unknown negatively charged counterion(s) must exist to balance the positive charge. The accumulation of large levels of cBPG in thermophilic and hyperthermophilic methanogens led to the hypothesis that cBPG could have a role as an enzyme thermoprotectant in these organisms. Later, it was reported that the potassium salt of cBPG (300 mmol/l) had a thermostabilizing effect at 90 °C on glyceraldehyde-3-phosphate dehydrogenase (GAPDH) and malate dehydrogenase (MDH) from *Mt. fervidus*. Other salts, such as sodium phosphate, potassium chloride and sodium chloride had little or no effect on the thermostability of these enzymes, but potassium phosphate also had a significant effect on enzyme stabilization [115]. On the other hand, cBPG did not appear to have any effect on the thermostabilization of rabbit GAPDH. The stabilizing effect of cBPG was also tested on purified DNA from *Mt. fervidus* and *Mt. sociabilis* which have a G + C ratio of only 33 mol%, and should therefore be intrinsically unstable at high growth temperatures. However, cBPG had no stabilizing effect on the purified DNA [115]. Recently, the potassium salt of cBPG was found to have a stabilizing effect, along with potassium sulfate and potassium phosphate, on the histone-like proteins of *Mt. fervidus* [120].

In other studies, cBPG was viewed as an intermediate of a novel branch of gluconeogenesis in *Methanobacterium thermoautotrophicum* ΔH [121, 122]. These authors showed that during the exponential phase of growth 2-phosphoglycerate (2-PG) and 2,3-bisphosphoglycerate (2,3-BPG), rather than cBPG, were the major intracellular solutes. A rapid interconversion between the three solutes was also observed and, during the exponential phase of growth the carbon flow was from 2-PG via 2,3-BPG to cBPG and to an unknown polymer, while in the stationary phase of growth the reverse was usually observed, leading to the conclusion that cBPG is an intermediate of an unusual branch of gluconeogenesis. Presumably, these authors did not consider the possibility that cBPG could also serve as a thermostabilizing agent. However, the possibility exists that cBPG could have more than one role, and the very high levels of this solute in hyperthermophilic methanogenic species indicates that this solute could play a role related to thermostabilization of cellular components.

2.3
Thermophiles and Hyperthermophiles

Many thermophilic and hyperthermophilic *Bacteria* and remove *Archaea* have been described in recent years [123, 124]. Our interest in these organisms stems from their evolutionary importance as representatives of ancient lineages of life, unusual cell components and metabolism, and because they possess thermostable enzymes and other molecules that can be exploited for industrial purposes. Many of these organisms originate from marine or saline hot springs, and are halotolerant or slightly halophilic; *Aquifex pyrophilus, Thermotoga maritima* and *Th. neapolitana, Thermus thermophilus* and *Rhodothermus marinus* are examples of slightly halophilic or halotolerant bacterial species that have high growth temperatures. Hyperthermophilic methanogenic and non-methanogenic species are very common within the *Archaea* and the majority of these also originate from marine geothermal areas [123].

Several novel low molecular weight solutes have recently been identified in thermophilic and hyperthermophilic bacteria and archaea which have not thus far been found in mesophilic species. For this reason there is a growing interest in their role as osmolytes and possible thermostabilizing agents that could be useful for the protection of enzymes, and other cell components and cells.

The polyol derivative di-*myo*-inositol-1,1'-phosphate (Fig. 7) was first identified in *Pyrococcus woesei* [125], although the same compound had already been detected but not identified in *Methanococcus igneus* [79,114]. Di-*myo*-inositol-1,1'-phosphate was subsequently identified in *Mc. igneus, Pyrococcus furiosus, Pyrodictium occultum*, and the hyperthermophilic bacteria *Thermotoga maritima* and *Thermotoga neapolitana* [27, 126–128]. In *P. furiosus*, this solute increases only slightly in concentration as the salinity of the medium is raised (Fig. 8), but the most dramatic increase in the levels of di-*myo*-inositol-1,1'–phosphate takes place as the temperature is raised above the optimum for growth [125–127]. This solute was also identified recently in *Pyrodictium occultum* in very high concentrations, although the relationship of its intracellular concentration with growth temperature and salinity has not been examined thus [128].

A role as a thermostabilizing agent was proposed for di-*myo*-inositol-1,1'-phosphate, and this solute was found to have a thermoprotective effect on glyceraldehyde-3-phosphate dehydrogenase from *P. woesei*. However, sodium citrate also stabilized this enzyme and the role of di-*myo*-inositol-1,1'-phosphate remains questionable [125]. It should be noted that solutes such as di-*myo*-inositol-1,1'-phosphate and cBPG may have a role as thermostabilizing agents in vivo, because other compounds cannot accumulate to high levels without exerting inhibitory effects on the cell and a role in the protection of cell components cannot, therefore, be discounted on the basis of in vitro results.

Di-*myo*-inositol-1,1'-phosphate was also recently detected in the slightly halophilic (1.5 – 5.0 % NaCl) and hyperthermophilic (optimum growth temperature, 80 °C) bacteria of the order Thermotogales *Thermotoga maritima* and *Th. neapolitana*. These species also accumulated two newly discovered solutes, identified as di-*myo*-inositol-1,3'-phosphate and di-mannosyl-di-*myo*-inositol-1,1'-phosphate (Fig. 7) [27]. An increase in the salinity of the growth medium resulted in a progressive increase in the concentration of all of the solutes, but was more pronounced with β-glutamate. However, di-*myo*-inositol-1,1'-phosphate was the major osmolyte at high salinities. The growth temperature also exerted a profound effect on the concentration of organic solutes in *Tt. neapolitana*. At the optimum salinity, the total pool of organic solutes increased in

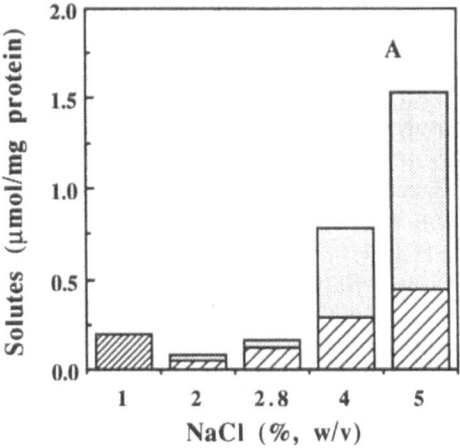

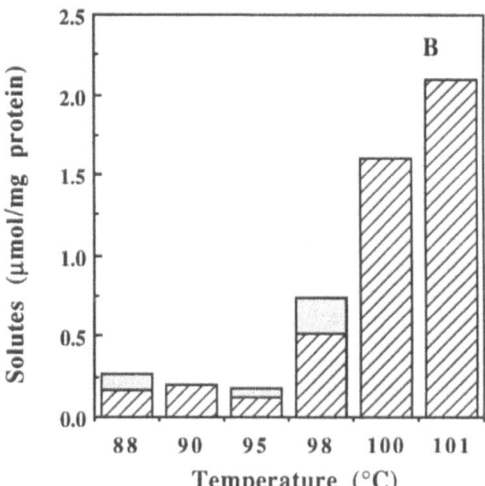

Fig. 8 A, B. Effect of: A NaCl concentration of the medium; B temperature on the accumulation of α-glutamate (▨), di-*myo*–inositol-1,1'-phosphate (▧) and 2-*O*-β-mannosylglycerate (■) by *Pyrococcus furiosus* during the late exponential phase of growth [169]. (With permission from the ASM)

concentration between 65 and 88 °C. The concentration of di-*myo*-inositol-1,1'-phosphate increased primarily between 65 and 80 °C, while di-*myo*-inositol-1,3'-phosphate and di-mannosyl-di-*myo*-inositol-phosphate, not detected during growth at the lowest temperature, increased dramatically as the growth temperature was raised above 80 °C. On the other hand, β-glutamate, the major solute at 65 °C, decreased to undetectable levels as the growth temperature was increased. The notion that some organic solutes have a role in the thermostabilization of macromolecules is reinforced once more by the behavior of di-*myo*-inositol-1,3'-phosphate and di-mannosyl-di-*myo*-inositol-phosphate (Fig. 9).

Unlike di-*myo*-inositol-1,1'-phosphate, which tends to increase in concentration concomitantly with the growth temperature, other compatible solutes appear to act as osmolytes in several thermophilic and hyperthermophilic organisms. The major compatible solutes responding to an increase in the NaCl concentration in *Mc. igneus* are β-glutamate and to a lesser extent α-glutamate [126]. In *P. furiosus*, however, the sugar derivative 2-*O*-β-mannosylglycerate is the primary osmolyte (Fig. 7) [127].

The α-isomer of mannosylglycerate (digeneaside) was initially detected in species of the red algal order *Ceramiales* [129]. Despite the low intracellular concentration of this solute it was believed, at first, to serve as a compatible solute in some of the species of the genera *Centroceras* and *Griffitsia* [130, 131]. However, recent results show that 2-*O*-α-mannosylglycerate does not serve as the primary osmolyte in the species *Caloglossa leprieurii*. In this species, as in many other algae, mannitol is the principal osmolyte [132]. However, digeneaside and *trans*-4-hydroxyproline betaine, present in low concentrations in *C. leprieuri*, are synthesized during the initial stages of osmoadaptation to higher external salt levels when Na^+, K^+ and Cl^- accumulate intracellularly, and before protecting levels of mannitol have been reached, leading to the proposal that digeneaside and *trans*-4-hydroxyproline betaine could protect enzyme function during the early stages of osmoadaptation [132].

The β-isomer of mannosylglycerate (Fig. 8) acts as an osmolyte in *P. furiosus* since, in contrast to di-*myo*-inositol-1,1'-phosphate, the concentration of this solute increases concomitantly with the NaCl concentration of the medium [127]. Mannosylglycerate is also the primary osmolyte in the slightly halophilic (0.5–7.0 % NaCl) thermophilic (optimum growth temperature, 65 °C) bacterium *Rhodothermus marinus* [133]. However, unlike *P. furiosus* which only accumulates the β-isomer and the red algae which only accumulate the α-isomer, *R. marinus* accumulates both isomers. At low salinities β-mannosylglycerate is the predominant compatible solute, but at the highest salinity (6 % NaCl) the α-isomer becomes the predominant osmolyte.

Trehalose and β-mannosylglycerate accumulate in the thermophilic and halotolerant bacterium *Thermus thermophilus* HB-8 concomitantly with the NaCl concentration of the growth medium, but trehalose was the most abundant solute at all salinities [133]. Trehalose was also the predominant organic solute in three other strains of *T. thermophilus*, two of which also accumulated very low amounts of betaine. The β-isomer of mannosylglycerate was also detected, along with roughly similar levels of trehalose and betaine in *Petrotoga miotherma*, a moderately thermophilic (optimum growth temperature, 55 °C) and slightly halophilic member of the bacterial order *Thermotogales* [27]. It also came as a surprise to identify β-mannosylglycerate in *Methanothermus fervidus*, because only c2,3-BPG had been detected previously [128]. The significance of β-mannosylglycerate in this thermophilic methanogen cannot be assessed at this time, but its presence in this organism strongly suggests that under some growth conditions it may serve as a compatible solute.

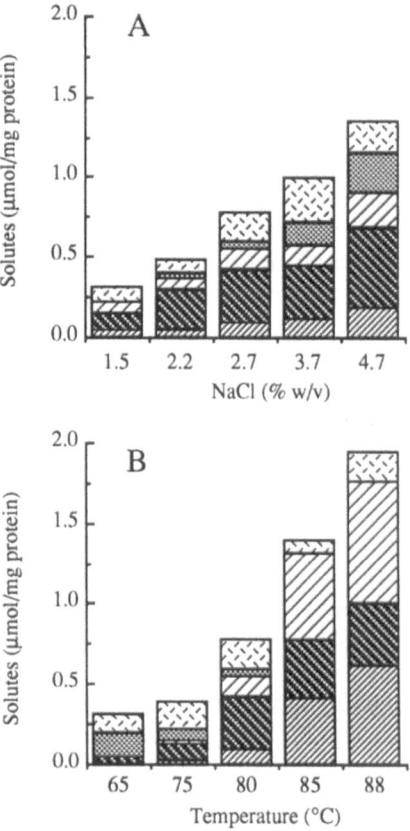

Fig. 9 A, B. Effect of: A NaCl concentration of the medium; B temperature on the accumulation of di-2-O-β-mannosyl-di-*myo*-inositol-1,1'-phosphate (▨), di-*myo*-inositol-1,1'-phosphate (●), di-*myo*-inositol-1,3'-phosphate (▨), α-glutamate (▨), and β-glutamate (▨) by *Thermotoga neapolitana* during the late exponential phase of growth [60]. (With permission from the ASM)

Another novel organic solute, identified as di-glycerol-phosphate has also been recently detected in the hyperthermophilic archaeon *Archaeoglobus fulgidus* [128]. This organism also accumulates α-glutamate, di-*myo*-inositol-1,1'-phosphate and vestigial levels of an unidentified isomer of di-*myo*-inositol-phosphate. Di-glycerol-phosphate appears to serve as the primary osmolyte in *A. fulgidus*, since α-glutamate is always a minor solute under conditions of high salinity and temperature. However, the concentration of di-glycerol-phosphate also increases as the growth temperature is raised above the optimum for growth, and reached levels similar to those of di-*myo*-inositol-1,1'-phosphate, which only increases in concentration concomitantly with the temperature. Di-glycerol-phosphate may, therefore, play multiple functions related to stress as has been proposed for trehalose.

As discussed above, trehalose appears to function as a general stress solute that may have a role, under unfavorable growth conditions, unrelated to osmotic stress. In view of the experimental data available thus far, it is difficult to envision trehalose as a primary compatible solute in many organisms subjected to osmotic stress. However, the accumulation of trehalose concomitantly with salinity in *T. thermophilus* indicates a role as an osmolyte and demands further experimentation. Trehalose has also been found to accumulate in several archaea of the order *Sulfolobales*, and in *Thermoplasma acidophilum* [128]. Since these organisms are not halotolerant trehalose probably serves a role unrelated to osmotic adjustment. It is also intriguing that trehalose was the only organic solute detected in the slightly halophilic archaeon *Pyrobaculum aerophilum*, but not in the closely related non-halotolerant species *Pyrobaculum islandicum*, both of which grow optimally at 95–98 °C. It should also be noted that other non-halotolerant thermophilic species, such as *Thermotoga thermarum* and *Fervidobacterium islandicum*, do not accumulate detectable levels of organic solutes [27] leading to the view putative thermostabilizing agents may not be necessary for protection of cell components against high temperatures. Compatible solutes found in thermophilic and hyperthermophilic organisms may, therefore, only have a role under osmotic stress. On the other hand, it is noteworthy that all of the non-halotolerant organisms were examined under optimum growth conditions, and some solutes may only become detectable at supraoptimal temperatures.

The medium used to grow these thermophilic and hyperthermophilic organisms contained yeast extract but, unexpectedly, betaine was only detected in low concentrations in two of four strains of *T. thermophilus* grown at 70 °C, and in higher levels in *P. miotherma* grown at 55 °C [27, 133]. These results indicate that one of the universal compatible solutes of mesophilic bacteria and methanogens does not serve its usual role in most of the thermophilic and hyperthermophilic organisms examined. It should be noted that these organisms are only slightly halophilic or halotolerant, and that the negatively charged osmolytes balancing the positive charge of potassium are probably adequate for osmoadaptation at low salt concentrations eliminating the need for neutral osmolytes such as betaine. Temperature may, to some extent, constrain the type of compatible solutes that accumulate in thermophiles and hyperthermophiles. Di-*myo*-inositol-1,1'-phosphate, di-*myo*-inositol-1,3'-phosphate, di-mannosyl-di-*myo*-inositol-phosphate, di-glycerol-phosphate and β-mannosylglycerate have not yet been found in mesophilic organisms, leading to the inevitable conclusion that some organic solutes are more suitable than others to protect cells and cell components at high growth temperatures. The existing data also indicate that mannosylglycerate is primarily involved in a response to increases in salinity, while the di-inositol-phosphate derivatives are involved in responses to high growth temperatures and could have roles in the thermostabilization of cellular components.

3
Osmolyte "Scavengers"

3.1
Source of Solutes

The massive developments of halophilic phototrophic bacteria (e.g. cyanobacterial mats) and other primary producers probably make compatible solutes the most abundant low molecular mass compounds in marine/saline/hypersaline ecosystems. Upon death and decay, but also through export mechanisms, such producer organisms release solutes into the environment, where they may be used by halophilic/halotolerant heterotrophic bacteria as a carbon source or as an easily accessible source of osmolytes, provided they have the necessary uptake systems [134, 135]. As a general rule, halophilic/halotolerant organisms usually prefer uptake of preformed osmolytes over de novo biosynthesis, hence the presence of external bacterial solutes in a saline environment has far reaching ecological effects and provides a selective advantage for rapid solute accumulators over other less efficient "scavengers". The phenomenon of solute uptake (solute "sponging") is even more vital for organisms which are unable to synthesize powerful compatible solutes themselves (e.g. *E. coli*, *Salmonella typhimurium*, *Rhizobium* sp.) and, in an environment of elevated salinity are, therefore, totally dependent on the external supply from de novo producers.

An important factor governing the supply of solutes into the surrounding medium is short term disturbance of medium osmolality through dilution due, for example, to sudden rainfall or local fresh water supplies. Many organisms unable to metabolize their own compatible solutes are totally dependent on extrusion systems, in order to master dramatic dilution events [136–139]. It is not surprising, therefore, to find that many, if not all, microorganisms able to thrive at lowered water activity have evolved membrane transport systems to recover leaked out solutes and to compete for the environmental osmolytes available.

3.2
Uptake Systems

Transport systems for betaine have, for example, been demonstrated and characterized in a large number of halophilic and halotolerant bacteria [140–145]. The K_m-data were either very low, as expected, for binding protein-dependent systems (approximately 1–10 µmol/l) or in a range more typical for low-affinity systems (approximately 50 µmol/l or above). The availability of glycine betaine (and often also ectoine) generally leads to rapid accumulation and at least partial replacement of other endogenic compatible solutes.

The most thoroughly studied organisms are, so far, the enterobacteria and rhizobia. Two distinct uptake systems, namely a low-affinity system ProP and a high-affinity system ProU, play an important role in this type of "acquired" osmoadaptation. ProP and ProU were originally associated with proline uptake (hence the name) and it was only later shown that they are in fact universal osmolyte transport systems, which accept a wide range of solutes [146–149]. Although the affinity for other powerful osmolytes such as hydroxyectoine and acetylated diamino acids has not yet been investigated, it is already known that the ProP/ProU system is relatively unspecific and accepts a whole range of similar compounds, from plant and animal sources (most importantly proline betaine, choline-O-sulfate and carnitine) (Fig. 10), and even man-made compounds like dimethylsulfonioacetate and 3-morpholino-1-propanesulfonic acid (MOPS), a common buffer component [93, 150].

Although both solute porters have a similar range of substrates, ProP and ProU are different and probably also serve different functions. The low affinity system ProP is characterized as a single polypeptide embedded in the cytoplasmic membrane and energized by the proton motive force. ProP is a constitutive system, one of the primary respondents to changes in cell turgor, and responsible for osmotically triggered rapid transport processes [151]. Despite its relatively low affinity, it probably accounts for most of the solute uptake from the medium [152]. The high affinity ProU, on the other hand, represents an inducible multi-component transport system [153], and it has been suggested that its primary function may be to recapture minute amounts of osmolytes which leak out from the cell [152]. For typical non-halophiles an elaborate solute scavenging system, like the above ProP/ProU, provides the only means for moderate osmotic adaptation and is, therefore, vital for survival in a low water environment.

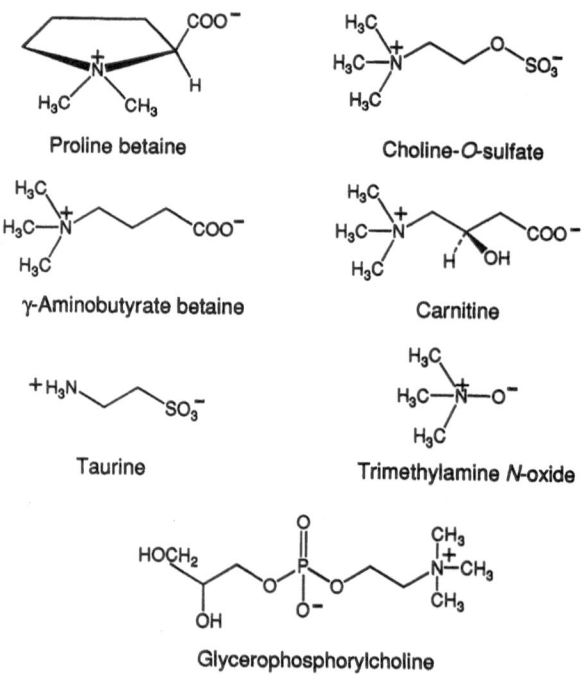

Fig. 10. Osmolytes of non-bacterial origin

Nevertheless, solute scavenging ("sponging") also has advantages for typical halophiles capable of de novo solute synthesis, because it enables these organisms to economize their biosynthetic expenditure and to modify their cytoplasmic solute pool depending on the range of external solutes available. Among the betaine producing phototrophic Proteobacteria efficient betaine transport systems have been characterized for *Aphanothece halophytica*, *Dactylococcopsis salina*, *Synechocystis* DUN52 and *Ectothiorhodospira halochloris* [140, 154, 155]. Of Gram-positive bacteria the halotolerant *Bacillus subtilis* and *Staphylococcus aureus* (both proline producers) have become model organisms for osmoadaptation at the molecular level [30, 141, 156–159].

3.3
Solute Precursors

The whole range of potential transport systems involved in osmoregulation may well stretch beyond current concepts. In addition, another class of "pre-osmolyte" transport systems has to be considered, which accumulate suitable precursors for compatible solute synthesis. The most important conversion mechanism sometimes still misnamed "betaine synthesis" is the oxidation of choline [160–161]. This common constituent of complex media components, which is usually derived from the hydrolysis of phosphatidylcholine, is accumulated via an independent uptake system and subsequently oxidized through the action of two enzymes: membrane bound O_2-dependent choline oxidase and NADP-dependent betaine aldehyde dehydrogenase [162–164]. The ability to oxidize choline is probably quite common in both Proteobacteria and Gram-positive bacteria [30, 157] and adds another dimension to solute "scavenging" and bacterial survival mechanisms.

4
Current Concepts of Compatible Solute Function

4.1
The Preferential Exclusion Model

The universal use of only a small number of compounds as osmotic regulators from bacteria to man has encouraged investigations into their underlying common principles [2, 6, 165]. Earlier attempts to correlate the stabilizing action with the physical parameter of surface tension were faced with the problem that some good stabilizers, such as betaine, contrary to expectations, decreased the surface tension while well known denaturants, such as urea, increased it (as did sugars and polyols). Attempts to describe enzyme stabilization in terms of increased surface tension of water-effecting forces of cohesion between water molecules have thus been unconvincing [166]. Subsequently, the main factor responsible for the stabilization effect of compatible solutes was disclosed to be preferential hydration and lack of interaction with proteins. Arakawa and Timasheff [166, 167] explain a solute's compatibility in the form of absence of interference, apparently not only in vitro, but also in the living cell. This is illustrated at the molecular level by the physicochemical principle of preferential exclusion of osmolytes from the immediate surface of proteins and other cytoplasmic structural elements [167], as demonstrated for a number of polyols, sugars, many amino acids and amino acid-type compatible solutes, but also for typical Hofmeister salts such as ammonium sulfate [168]. The phenomenon of preferential hydration of protein favors a more compact protein conformation, opposes an increase in surface area of the protein and, since unfolding usually results in a surface area increase, favors the native state.

4.2
Water Structure Implications

The authors in favor of the preferential exclusion model provided ample support for their stabilization concept through experimental data obtained from dialysis experiments. The question as to why these solutes were excluded form the surface of a protein, however, still remained unsolved. It was the work of Philippa Wiggins [169] on water structure in gels and at the interface of hydrophobic and charged polymers which helped to explain the preferential exclusion of stabilizing solutes. Because water at interfaces, "hydration water", is structurally different (more dense in the case of charged polymers such as proteins), exclusion of compatible solutes is seen as a preference for the non-hydration water, which is characterized by its lower density and stronger hydrogen bonds. This decrease in density is caused by an osmolality gradient of counterions (high concentration near the surface of a charged polymer) and is seen as a means of lowering the local chemical potential to equal that of hydration water [169].

The distinction between different fractions of water of different density (and in a dynamic equilibrium with each other) is experimentally difficult to prove, but this investigator has accumulated a wealth of information on model systems such as dextran sulfate solutions, cellulose acetate films, gels of the Biogel P series and ion exchange resins. These experiments present evidence for the existence, albeit transient, of high and low density water depending on the nature of the interface [170–174].

Since changes in the density of water are accompanied by changes in water hydrogen bond strength, and subsequently also in physical and chemical properties of the liquid, different solvent properties of co-existing water populations may result in an uneven distribution of solutes. Philippa Wiggins and co-workers were able to demonstrate that univalent anions (HCO_3^-, $H_2PO_4^-$, Cl^-) and large univalent cations (NH_4^+, K^+) as well as compatible solutes are excluded from the dense hydration water of charged surfaces whereas small highly hydrated cations ($Mg^{2+} > Ca^{2+} > H^+ > Li^+ > Na^+$), small anions with high surface charge (e.g. HPO_4^{2-}) and hydrophobic molecules have a tendency to accumulate in this water fraction. These "hydration water" solutes have a negative entropy of hydration, because the order they impose on neighboring water molecules is greater than the disorder generated in the transient layer (structure makers). Electrolytes excluded from the hydration shell of charged polymers, on the other hand, are usually powerful structure breakers, i.e. they generate relatively more disorder than order. As a consequence of their accumulation into low density water, they convert it back to normal structure and solvent properties, unless the concentration is very low and, therefore, destroy their own preferred environment. This accounts for the difficulty in demonstrating low density water (stretched water) using probe molecules. Due to their unique properties compatible solutes also accumulate into structured water, but without breaking down its structure [171]. This behavior is probably caused by their hydrophobic moieties, which induce stretched water around them and – in doing so – protect it from collapse.

4.3
Towards an Understanding of the Stabilization Phenomenon

4.3.1
Aqueous Solutions (In Vitro Systems)

It has been demonstrated recently by NIR measurements [175] that compatible solutes such as betaine, proline and ectoine enhance the formation of structured water in their vicinity. As pointed out by the same authors, methodological restrictions only allow data to be obtained at concentrations of 1 mol/l and higher. Still there seems to be a general tendency that this structure forming effect is even more pronounced at dilute concentrations. These findings are, therefore, consistent with the view that compatible solutes are not only excluded from enzyme hydration shells [167, 176] but also enhance a structured environment in bulk by enforcing the formation of hydration clusters, a property which had already been proposed earlier [177].

As has been pointed out by Arakawa and Timasheff [166], to understand protein stabilization, both the preferential exclusion effect and the solute's tendency for binding to exposed protein surface have to be considered. From this point of view, betaine with three methyl groups should have a greater affinity for non-polar groups than glycine. This molecular characteristic might compete with, or at least attenuate, the preferential hydration effect. This would explain the rank order of glycine-based osmolytes with respect to T_m and also help to understand why, in this particular example, the stabilizing effect of betaine and dimethylglycine was reversed at very high concentrations (4 mol/l for betaine and 6 mol/l for dimethylglycine). Even if one assumes that the exclusion effect should be the same for globular proteins of a similar size and, as such, cause similar stabilization, a solute's affinity towards the surface is clearly dependent on the nature of the protein, and the two opposing forces will be of different magnitudes with different proteins. This would clearly explain largely different effects depending on the solute and the protein in question, and also destroy the illusion that a specific compatible solute will always be a stabilizer, irrespective of the enzyme in question. A more holistic approach of stabilization effects considering the specific interaction of enzyme, solute and water will more likely help to explain the vastly diverging effects of solutes under different test conditions.

4.3.2
Cells and Cell Model Systems

Experimental conditions in simple solution are far from conditions in the living cell. Good model systems are, for example, concentrated dextran sulfate solutions, where water is probably either compacted or stretched (containing less than 3 g/g dry weight of water). This probably provides a good approximation of cells, which have a water content of 3–4 g water per g dry weight. Such highly condensed systems are sensitive to solute concentrations in the low mmol/l range, and it has been shown that betaine enhanced the viscosity dramatically at sub-molar concentrations [174]. In ion exchange column experiments betaine was also shown to protect low density water, but in a remarkably non-linear manner (maximum effect at 50 mmol/l). This seemingly unpredictable and strongly concentration-dependent behavior is explained by a number of partially opposing effects, which have to be considered individually and in combination: the preference of compatible solutes for less dense water, the stabilization of less dense water (e.g. via methyl groups) at low concentrations, and a counteracting effect with increasing concentrations, as the osmolality gradient between high density water and less dense water is diminished, allowing both water populations to return to normal density.

This has important consequences for the living cell, where the intracellular environment is probably comparable to a gel. The enzymatic machinery of a living cell experiences a densely packed microenvironment with anionic polymers such as proteins and nucleic acids and close vicinity of negatively charged phospholipids and potassium counterions (approx. 0.5 µmol/mg protein). The cytoplasm, therefore, behaves like a concentrated solution of potassium salts of polyanions, in which additional solutes balance the increasing osmolarity of the medium [20]. Most of the cytoplasmic water is, therefore, probably structured: dense water in hydration zones and less dense water in "bulk". Exposure to a low water environment caused by increased salinity would lead to rapid efflux of water and concomitant changes in cytoplasmic volume, concentration of solutes and, consequently, also water structure. As high potassium concentrations, the cell's primary osmoresponse, are known to have detrimental effects in vitro inhibiting the activity of many enzymes [6, 178, 181], it was first assumed that the accumulation of potassium was responsible for the limitation of growth and that compatible solutes, by replacing potassium, relieved this stress situation. An investigation into the interconnected parameters of volume, solute concentration and growth rate performed by Cayley and coworkers [39], however, revealed a number of interesting observations in osmoadapting E. coli, showing that cytoplasmic volume and growth rate were interrelated. They concluded that retarded biological functions are possibly explained by molecular crowding and concomitantly reduced diffusion rates of proteins and metabolites [179] or by non-specific aggregation of proteins driven by the same effect. Although the precise magnitude of their data has been questioned later by others [180], evidence remains that shrinking may be directly connected with growth inhibition. In the view of these results, water loss from a gel-like system would cause increased accumulation of electrolytes in hydration water and – according to the Wiggins model – a dramatic expansion of the remaining non-hydration water, strengthening of hydrogen bonding and the subsequent slow down of metabolic processes. The accumulation of potassium glutamate, the cells' primary, and in many cases only, response, seemed to suffice for low level adaptation only with little effect on volume, while the presence of compatible solutes such as betaine largely restored the original volume, even though the total amount of solutes within the cell remained almost unchanged [39]. The authors concluded that betaine which replaces potassium, glutamate (and trehalose) must have a significantly higher osmotic coefficient. In view of the structure forming (kosmotropic) properties of betaine it would be equally justified to conclude that this compatible solute, in contrast to other less "compatible" solutes, restores the original water structure in bulk at an overall reduced water activity. Hence the primary function of compatible solutes seems to be osmotic, at first sight only, providing equilibrium across the membrane. In addition, compatible solutes appear to be special because of unique structural features which enable accumulation of these solutes in non-hydration water and the ability to restore the original volume and balance of structured water.

5
Applications as Stress Protectants

Low molecular mass solutes had already been used empirically for the stabilization of native enzymes and other biomolecules, long before the term "compatible solute" was coined. Most investigations were, at the time, concerned with the protective effect of sugars, polyols (primarily glycerol), proline and glycine betaine against heat stress and freeze thaw treatment of enzymes, cells and cell organelles [181–186]. Since the original definition of compatible solutes by Brown [5], compatibility of osmolytes with enzyme and other cellular functions in vivo and in vitro has been sufficiently demonstrated. From our present concept of compatible solutes' function in nature (stabilization in a low water environment) we would expect a pronounced effect on enzymes, DNA, membranes and even whole cells in situations where water is withdrawn from the system. This would apply to freezing and drying, and other denaturation processes such as heat, if one assumes a common underlying physical phenomenon, involving, for example, stabilization of a functional hydration shell. Without doubt such general stress protectants would find a wide range of potential applications in bio- and enzyme technology.

5.1
High Temperature

Protection of enzymes against heat inactivation with polyols, amino acids and betaines had been observed many times, but markedly different effects that depend on experimental conditions and the enzyme under investigation provided obstacles to our understanding of the underlying principles [187–190]. Putative anionic temperature stabilizers from hyperthermophiles such as cyclic 2,3-bisphosphoglycerate and di-myo-inositol-1,1'-phosphate have also been employed for similar studies. However, the temperature stabilization exerted by this special protectant was not always consistent and was sometimes rivaled by "normal" inorganic Hofmeister salts such as potassium sulfate [115, 120, 125].

The most reliable methods for a quantitative assessment of a solute's temperature stabilizing power in solution are probably differential spectrophotometry and differential scanning calorimetry (DSC) with fully reversible model systems such as ovalbumine, ribonuclease A and egg white lysozyme. When comparing the effect of a whole range of polyols on thermal denaturation of lysozyme with DSC, a general tendency to enhance thermal stability was demonstrated. The melting temperature (T_m) increased with increasing polyol concentration and with the number of hydroxyl groups per molecule [191, 192, 193]. Comparable results were obtained using differential spectrophotometry with chymotrypsinogen (293 nm) and ribonuclease A (287 nm) in the presence of glycerol, where T_m of the RNase increased by 7 °C in the presence of 40 vol. % glycerol [177]. Santoro and coworkers [191], again using DSC methods, have shown even higher effects of temperature stabilization of RNase A in the presence of glycine and its methylated derivatives (sarcosine, dimethylglycine, and betaine) reaching a maximum value of ΔT_m of 22 °C for sarcosine. The fact that a temperature stabilization of similar magnitude is hard to achieve even with the most sophisticated strategies of molecular design, underlines once again the importance of low molecular weight compounds as powerful solvent modifiers with a large biotechnological potential.

5.2
Low Temperature and Freezing

A solute's potential effect on low temperature stability has been largely unexplored because of experimental difficulties due to freezing at sub-zero temperatures. There is only a small amount of experimental data available. The cold labile pyruvate: Pi dikinase from *Zea mays* (a C4 plant), which undergoes dissociation into monomers at 0 °C [194] was cold-stabilized by solutes such as glycerol, proline, betaine and trimethylamine oxide (TMAO) and the view was presented that the stabilization of the native oligomeric conformation was caused by modifications of physical solvent properties. For the majority of enzymes, however, low temperature effects coincide with freezing, which is usually associated with a number of additional stress factors. As aqueous solutions of biomolecules are cooled down, water freezes out as ice and salts are concentrated and crystallize at various temperatures depending on their respective solubilities. The lowest temperature at which the frozen state and the liquid state can exist in equilibrium is called the eutectic point. At this stage pH and solute composition of the eutectic mixture can be very different from that at room temperature [195]. Thus a protein/biomolecule is subjected to a variety of changes in physical conditions during freezing: temperature, pH, ionic strength, and water activity. The influence of individual parameters is, therefore, difficult to resolve. For oligomeric proteins, dissociation into subunits is often observed during freezing. On the other hand, physical contact between molecules (when water is removed) may also promote irreversible aggregation. Common test enzymes used for the assessment of solute cryoprotection are lactate dehydrogenase (LDH, rabbit muscle, M_4 isoenzyme, type V-S) and phosphofructokinase (rabbit muscle, type II). The results obtained with a whole range of solutes applied during freeze-thawing of LDH in potassium phosphate buffer showed remarkably similar effects and concentration dependent protection (usually between 60 and 80% residual activity after one freeze-thaw cycle) [190, 194].

Generally, most solutes from the most diverse classes of compounds displayed a saturation effect between 0.5 and 1.0 mol/l concentration (glucose, lactose, sucrose, mannitol, sorbitol, xylitol, ethylene glycol, proline, γ-aminobutyric acid, sarcosine, sodium glutamate, sodium acetate, lysine-HCl) [196]. Surprisingly the stabilization in the presence of betaine (40%) and glycine (20%) was markedly lower than with sarcosine (60%) and at 1 mol/l was still far from its point of saturation. Whereas TMAO proved to be most efficient (60% protection) at a concentration of only 100 mmol/l, best protection with almost no loss of activity at 25 mmol/l was, however, achieved with polyethylene glycol (MW 600) [196]. Under comparable test conditions the stabilization of ectoine in a freeze-thaw situation proved very similar to that of other osmolytes, while that of hydroxyectoine was shown to be exceptional (100% residual activity over four freeze thaw cycles) [190]. The stabilization conferred to phosphofructokinase (PFK) was comparable in some aspects, but differed in others. Betaine, for example, proved to be a much better stabilizer for this enzyme, while glucose, glycerol and inositol were relatively ineffective [190, 197]. It is important to note that the stabilizing effect of compatible solutes on phosphofructokinase (PFK) was greatly enhanced by the addition of very small amounts of $ZnSO_4$ (and Ni, Cu, Co), enabling maximum protection at solute concentrations as low as 50 mmol/l or less. The mechanism by which cations alter the cryoprotective capacity of organic solutes, at least with PFK is presently not known [197] and adds yet another complexity to the puzzle of how labile proteins are preserved.

5.3
Desiccation

The observation that a number of excellent freeze protectants did not perform well as stabilizers during drying (freeze-drying), while others, especially sugars, gave best results [198–200], demonstrated that freezing and drying are principally different stress factors. By removing the so-called non-freezable water from the hydration shell it becomes essential to substitute hydration water at critical places [79, 200, 201]. Hence, it was suggested that the capacity of sugars to preserve dried enzymes depends on binding to the protein during the final stages of dehydration. The importance of OH-groups is, for example, demonstrated by comparing the protection of ectoine and hydroxyectoine. While the former exerts little protection (comparable to that of betaine) for both PFK and LDH, the latter proved to be as good a stabilizer as trehalose [190].

On the basis of infrared (IR) spectroscopy and solid state nuclear magnetic resonance spectroscopy (NMR) it was, in fact, shown that hydroxy groups of solutes such as trehalose (in the dry state) interact with polar regions of the enzyme simulating the effect of water [48, 201] and that, for example, trehalose maintains the integrity of the membrane upon dehydration by binding to the headgroups of phospholipid bilayers, which in turn ensures a liquid crystalline-like phase in the dry state [47, 167, 202–204]. It seems plausible that, in a living system, stabilization of the most sensitive target (the membrane) is vital for survival of unhydrobiotic organisms and for preservation and shelf life of commercial microbial cultures. As a practical consequence of the "water replacement" hypothesis [204], hydroxyl group carriers such as trehalose have important implications. Experiments exploring the effect of compatible solutes on the preservation of freeze-dried *E. coli* cells have confirmed the importance of hydroxyl groups, not only in the case of sugars, but also for hydroxyectoine [49]. Stabilization of phosphofructokinase (PFK) during air- and freeze-drying in the presence of sugars was also dramatically improved by transition metals (such as Zn). The possible contribution of metal ions cannot, therefore, be ignored in future stabilization studies [198, 199].

5.4
Denaturing Solutes

Most investigations on a possible salt protective effect on enzymes were performed with betaine, but also with proline and glycerol, and were able to demonstrate that for most, but not all, cases salt tolerance can indeed be gained by modifying the enzymes' microenvironment [183, 185, 186, 205, 206]. If the observed relief of salt inhibition can be explained by a general stabilizing effect of compatible solutes which counteracts all kinds of unfolding/denaturation processes, then the presence of compatible solutes should also compensate for deleterious destabilizing effects of denaturants such as urea. This has, in fact, been demonstrated for a number of cases, and compared to the natural system of osmoadaptation in elasmobranch fish, where both urea and betaine/TMAO are used as an osmolyte mixture, one component compensating the deleterious effect of the other [6, 92, 165, 207]. In view of the complexity of water-solute-protein interactions, in depth analysis of the underlying molecular principles is slow and still in progress, and we have now come to a point where we envision the whole dimension of the problem, and realize that the puzzle probably has more pieces than we have yet identified.

6
Conclusions

Studying the organisms which have adapted to conditions of low water activity, dehydration, high and low temperature (or other environments of extremophi-·lic nature) provides an insight into how nature has approached the problem of maintaining a functional cell under stress conditions. As has been described here, organisms adapting to extreme conditions produce and accumulate certain solutes which are believed to confer protection without affecting the function of the living cell. The increase in stability obtained by this simple strategy (for the few cases studied) was shown to rival some of the best protein design strategies. Since these solutes need to stabilize a great number of enzymes and cell components, it is reasonable to assume that this protective effect is of a very general nature and may be exploited for biotechnological purposes for the stabilization of biological structures, be they enzymes, proteins, DNA, membranes or even whole cells. A clear and complete understanding of just how this stabilization is accomplished and whether it can be predicted for specific enzyme solute combinations remains a point for future research.

Acknowledgments. The work in the author's laboratories is supported by the European Community Biotech Programme (Extremophiles as Cell Factories) Contract No. BIO4-CT96-0488, and by Praxis XXI and FEDER Programmes (Praxis 2/2.1/BIO/20/94), Portugal to H.S and M.C.

Addendum added in proof. Several studies related to the subject of this review appeared recently that warrant mention. Genes for the synthesis of ectoine by *Marinococcus halophilus* were characterized and expressed in *E. coli* [Louis P, Galinski EA (1997) Microbiol 143:1141]. β-Mannosylglycerate was shown to protect several enzymes from thermal denaturation and freeze-drying [Ramos A, Raven NDH, Sharp RJ, Bartolucci S, Rossi M, Cannio R, Lebbink J, Van der Oost, de Vos WM, Santos H (1997) Appl Environ Microbiol 63:4020]. The compatible solute sulfotrehalose, was identified in extremely alkaliphilic archaea [Desmarais D, Jablonski PE, Fedarko NS, Roberts MF (1997) J Bacteriol 179:3146].

7
References

1. Brown AD (1990) Microbial water stress physiology. Principals and perspectives. Wiley, Chichester
2. Galinski EA (1995) Adv Microbial Physiol 37:273
3. Brown AD (1978) Adv Microbial Physiol 17:181
4. Robinson RA, Stokes RH Electrolyte solutions, 2nd edn. Butterworths, London
5. Brown AD (1976) Bacteriol Rev 40:803
6. Yancey PH, Clark ME, Hand SC, Bowlus RD, Somero GN (1982) Science 217:1214
7. Reed RH (1984) Plant Cell and Environ 7:165
8. Kushner DJ (1988) Can J Microbiol 34:482
9. Ventosa A (1989) Taxonomy of halophilic bacteria. In: da Costa MS, Duarte JC, Williams RAD (eds) Microbiology of extreme environments and its potential for biotechnology. Elsevier, London, p 262
10. da Costa MS, Nobre MF (1989) Polyol accumulation in yeasts in response to water stress. In:da Costa MS, Duarte JC, Williams RAD (eds) Microbiology of extreme environments and its potential for biotechnology. Elsevier, London, p 310
11. Galinski EA, Trüper HG (1994) FEMS Microbiol Rev 15:95
12. Eisenberg H, Maverech M, Zaccai G (1992) Adv Protein Chem 43:1
13. Oren A (1985) Can J Microbiol (1985) 32:4
14. Rengpipat S, Lowe SE, Zeikus JC (1988) J Bacteriol 170:3065
15. Oren A, Gurevich P (1993) FEMS Microbiol Letters 108:287
16. Avron M (1986) Trends Biochem Sci 11:5
17. Ben-Amotz A, Avron M (1973) Plant Physiol 51:875
18. Epstein W, Schultz SG (1965) J Gen Physiol 49:221
19. Epstein W (1986) FEMS Microbiol Rev 39:73
20. Cayley S, Lewis BA, Guttman HJ, Record MT Jr (1991) J Mol Biol 222:281
21. McLaggan D, Logan TM, Lynn DG, Epstein W (1990) J Bacteriol 172:3631
22. Tempest DW, Meers JL, Brown CM (1970) J Gen Microbiol 64:171
23. Dinnbier U, Limpinsel E, Schmid R, Bakker EP (1988) Arch Microbiol 150:348
24. Welsh DT, Reed RH, Herbert RA (1991) J Gen Microbiol 137:745
25. Henrichs SM, Cuhel R (1985) Appl Environ Microbiol 50:543

26. Robertson DE, Lesage S, Roberts MF (1990) Biochim Biophys Acta 992:320
27. Martins LO, Carreto LS, da Costa MS, Santos H (1996) J Bacteriol (in press)
28. Mackay MA, Norton RS, Borowitzka LJ (1984) J Gen Microbiol 130:2177
29. Nagata S, Ogawa Y, Mimura H (1991) J Gen Appl Microbiol 37:403
30. Graham JE, Wilkinson BJ (1992) J Bacteriol 174:2711
31. Whatmore AM, Chudek JA, Reed RH (1990) Microbiology 136:2527
32. Whatmore AM, Reed RH (1990) J Gen Microbiol 136:2521
33. Reed RH, Richardson DL, Warr SRC, Stewart WDP (1984) J Gen Microbiol 130:1
34. Welsh DT, Herbert RA (1993) FEMS Microbiol Ecol 13:145
35. D'Souza-Ault MR, Smith LT, Smith GM (1993) Appl Environ Microbiol 59:473
36. Madkour MA, Smith LT, Smith GM (1990) Appl Environ Microbiol 56:2876
37. Smith LT, Smith GM (1989) J Bacteriol 171:4714
38. Larsen PI, Sydnes LK, Landfald B, Strøm AR (1987) Arch Microbiol 147:1
39. Cayley S, Lewis BA, Record MT Jr (1992) J Bacteriol 174:1586
40. Strøm AR, Kaasen I (1993) Mol Microbiol 8:205
41. Van Laere A (1989) FEMS Microbiol Rev 63:201
42. Wiemken A (1990) Antonie van Leeuwenhoek 58:209
43. Hengge-Aronis R (1993) Cell 72:165
44. Gadd GM, Chalmers K, Reed RH (1987) FEMS Microbiol Lett 48:249
45. Donnini C, Puglisi PP, Vecli A, Marmiroli N (1988) J Bacteriol 170:3789
46. Hino A, Mihara K, Nakashima K, Takano H (1990) Appl Envrion Microbiol 56:1386
47. Crowe JH, Crowe LM, Carpenter JF, Wistrom CA (1987) Biochem J 242:1
48. Carpenter JF, Crowe JH (1989) Biochem. 28:3916
49. Louis P, Trüper HG, Galinski EA (1994) Appl Microbiol Biotechnol 41:684
50. Severin J, Wohlfarth A, Galinski EA (1992) J Gen Microbiol 138:1629
51. Smith LT, Smith GM, Madkour MA (1990) J Bacteriol 172:6849
52. Borowitzka LJ, Demmerle S, Mackay MA, Norton RS (1980) Science 210:650
53. Reed RH, Chudek JA, Foster R, Stewart WDP (1984) Arch Microbiol 138:333
54. Galinski EA, Trüper HG (1982) FEMS Microbiol Lett 13:357
55. Galinski EA, Pfeiffer HP, Trüper HG (1985) Eur J Biochem 149:135
56. Galinski EA (1986) PhD thesis, University of Bonn
57. Trüper HG, Galinski EA (1986) Experientia 42:1182
58. Trüper HG, Galinski EA (1990) FEMS Microbiol Rev 75:247
59. Wohlfarth A, Severin J, Galinski EA (1990) J Gen Microbiol 136:705
60. Wohlfarth A, Severin J, Galinski EA (1993) Appl Microbiol Biotechnol 39:568
61. del Moral A, Severin J, Ramos Cormenzana A, Trüper HG, Galinski EA (1994) FEMS
    Microbiol Lett 122:165
62. Gouesbet G, Blanco C, Hamelin J, Bernard T (1992) J Gen Microbiol 138:959
63. Bernard T, Jebbar M, Rassouli Y, Himdi-Kabbab S, Hamelin J, Blanco C (1993) J Gen Micro-
    biol 139:129
64. Frings E, Kunte HJ, Galinski EA (1993) FEMS Microbiol Lett 109:25
65. Schmitz RPH, Galinski EA (1996) FEMS Microbiol Lett 142:195
66. Storey R, Wyn Jones RG (1977) Phytochem 16:447
67. Hocking AD, Norton RS (1983) J Gen Microbiol 129:1915
68. Loos H, Krämer R, Sahm H, Sprenger GA (1994) J Bacteriol 176:7688
69. Kets EP, Galinski EA, de Wit M, de Bont JAM, Heipieper HJ (1996) J Bacteriol 178:6665
70. Impellizzeri G, Mangiafico S, Oriente G, Piattelli M, Sciuot S, Fattorusso E, Magno S,
    Santacroce C, Sica D (1975) Phytochem:14:1549
71. Kauss H (1979) Prog Phytochem 5:1
72. Mackay MA, Norton RS, Borowitzka LJ (1983) Mar Biol 73:301
73. Kremer BP, Kirst GO (1981) Plant Sci Lett 23:349
74. Hagemann M, Erdmann N (1994) Microbiology 140:1427
75. Joset F, Jeanjean R, Hagemann M (1996) Physiol Plant 96:738
76. Hagemann M, Schoor A, Erdmann N (1996) J Plant Physiol (in press)
77. Pocard JA, Smith LT, Smith GM, Le Rudulier D (1994) J Bacteriol 176:6877

78. Killham K, Firestone MK (1984) Appl Environ Microbiol 47:301
79. Robertson D, Noll D, Roberts MF, Menaia J, Boone RD (1990) Appl Environ Microbiol 56:563
80. Lai MC, Sowers KR, Robertson DE, Roberts MF, Gunsalus RP (1991). J Bacteriol 173:5352
81. Roberts MF, Lai MC, Gunsalus RP (1992) J Bacteriol 174:6688
82. Galinski EA, Oren A (1991) Eur J Biochem 198:593
83. Khunajakr N, Shinmyo A, Takano M (1989) Compatible solutes in a halotolerant bacterium. Annual Reports of International Centre of Cooperative Research in Biotechnology, Osaka, Japan 12:157
84. Peters P, Galinski EA, Trüper HG (1990) FEMS Microbiol Lett 71:157
85. Tao T, Yasuda N, Ono H, Shinmyo A, Takano M (1992) Purification and characterization of 2,4-diaminobutyric acid transaminase from *Halomonas* sp. Annual Reports of International Centre of Cooperative Research in Biotechnology. Osaka University, Japan 15:187–199
86. Inbar L, Lapidot A (1988) J Biol Chem 263:16014
87. Galinski EA, Sauer T, Trüper HG (1994) Verfahren zur in vivo Gewinnung von Inhaltsstoffen aus Zellen. OS DE 42 44 580 A1, Appl. P 4244580.9 (Cl. C12 P1/00) 31 Dec 92
88. Frings E, Sauer T, Galinski EA (1995) J Biotechnol 43:53
89. Wyn Jones RG, Storey R (1981) Betaines. In: Paleg LG, Aspinall D (eds) Physiology and biochemistry of drought resistance in plants. Academic Press, Sydney
90. Blunden G, Gordon SM, Crabb TA, Roch OG, Rowan MG, Wood B (1986) Magnetic Resonance in Chemistry 24:965
91. Larher F (1988) Plant Physiol Biochem 26:35
92. Bagnasco S, Balaban R, Fales HM, Yang YM, Burg M (1986) J Biol Chem 261:5872
93. Mason TG, Blunden G (1989) Bot Mar 32:313
94. Rhodes D, Hanson AD (1993) Ann Rev Plant Physiol and Plant Mol Biol 44:357
95. Dragolovich J (1994) J Exp Zool 268:139
96. Coughlan SJ, Wyn Jones RG (1982) Planta 154:6
97. Tschichholz I (1994) PhD Thesis, University Bonn
98. Robertson DE, Lai MC, Gunsalus RP, Roberts MF (1992) Appl Environ Microbiol 58:2438
99. Imhoff JF, Rodriguez-Valera F (1984) J Bacteriol 160:478
100. Imhoff JF (1986) FEMS Microbiol Rev 39:57
101. Greene RC (1962) J Biol Chem 237:2251
102. Maw GA (1981) The biochemistry of sulfonium salts In: Stirling CJM, Patai S (eds) The chemistry of the sulfonium group, pt 2. Wiley, Chichester, UK, p 703
103. Mathrani IM, Boone DR (1985) Appl Environ Microbiol 50:140 143
104. Paterek JR, Smith PH (1985) Appl Environ Microbiol 50:877
105. Zhilina TN (1986) Syst Appl Microbiol 7:216
106. Zhilina TN, Zavarzin GA (1990) FEMS Microbiol Rev 87:315
107. Menaia JAGF, Duarte JC, Boone DR (1993) Experientia 49:1047
108. Sowers KR, Robertson DE, Noll D, Gunsalus RP, Roberts MF (1990) Proc Natl Acad Sci USA 87:9083
109. Sprott GD, Jarrell (1981) Can J Microbiol 27:444
110. Ciulla R, Clougherty C, Belay N, Krishnan S, Zhou C, Byrd D, Roberts MF (1994) J Bacteriol 176:3177
111. Gorkovenko A, Roberts MF, White RH (1994) Appl Environ Microbiol 60:1249
112. Sowers KR, Gunsalus RP (1995) Appl Environ Microbiol 61:4382
113. Lai M, Gunsalus RP (1992) J Bacteriol 174:7474
114. Robertson DE, Roberts MF (1991) Biofactors 3:1
115. Hensel R, König H (1988) FEMS Microbiol Lett 49:75
116. Kanodia S, Roberts MF (1983) Proc Natl Acad Sci USA 80:5217
117. Seeley RJ, Farney DE (1983) J Biol Chem 258:10835
118. Tolman CJ, Kanodia S, Daniels L, Roberts MF (1986) Biochim Biophys Acta 886:345
119. Kurr M, Huber R, König H, Jannasch HJ, Fricke H, Trincone A, Kristjansson JK, Stetter KO (1991) Arch Microbiol 156:239

120. Reeve JN, Grayling RA, Pereira S, Li W-TL, Priestley ND, Sandman K (1996) Histones and chromatin structure in hyperthermophiles. Proceedings of the First International Congress on Extremophiles, p 6
121. Gorkovenko A, Roberts MF (1993) J. Bacteriol 175:4087
122. Sastry MVK, Robertson DE, Moynihan JA, Roberts MF (1992) Biochemistry 31:2926
123. Blöchl E, Burggraf S, Fiala G, Lauerer G, Huber G, Huber R, Rachel R, Segerer A, Stetter KO, Völkl P (1995) World J Microbiol Biotechnol 11:9
124. Kristjansson JK, Stetter KO (1992) Thermophilic bacteria. In Kristjansson JK (ed) Thermophilic bacteia. CRC Press, Boca Raton, p 1
125. Scholz S, Sonnenbichler J, Schafer W, Hensel R (1992) FEBS Lett 306:239
126. Ciulla RA, Burggraf S, Stetter KO, Roberts MF (1994) Appl Environ Microbiol 60:3660
127. Martins LO, Santos H (1995) Appl Environ Microbiol 61:3299
128. Martins LO, Huber R, Huber H, Stetter KO, da Costa MS, Santos H (1997) Appl Environ Microbiol 63:896
129. Kremer BP, Vogl R (1975) Phytochem 14:1309
130. Bisson MA, Kirst GO (1979) Aust J Plant Sci 6:523
131. Kirst GO, Bisson MA (1979) Aust J Plant Sci 6:539
132. Karsten U, Barrow KD, Mostaert AS, King RJ, West JA (1994) Plant Physiol Biochem 32:669
133. Nunes OC, Manaia CM, da Costa MS, Santos H (1995) Appl Environ Microbiol 61:2351
134. Gauthier MJ, LeRudulier D (1990) Appl Environ Microbiol 56:2915
135. Ghoul M, Bernard T, Cormier M (1990) Appl Environ Microbiol 56:551
136. Reed RH, Warr SRC, Kerby NW, Stewart WDP (1986) Enz Microbiol Technol 8:101
137. Fulda S, Hagemann M, Libbert E (1989) Arch Microbiol 153:405
138. Galinski EA, Herzog RM (1990) Arch Microbiol 153:607
139. Tschichholz I, Trüper HG (1990) FEMS Microbiol Ecol 73:181
140. Moore DJ, Reed RH, Stewart WDP (1987) Arch Microbiol 147:399
141. Molenar D, Hagting A, Alkema H, Driessen AJM, Konings WL (1993) J Bacteriol 175:5438
142. Bae JH, Anderson SH, Miller KJ (1993) Appl Environ Microbiol 59:2734
143. Ko R, Smith LT, Smith GM (1994) J Bacteriol 176:426
144. Kempf B, Bremer E (1995) J Biol Chem 270:16701
145. Farwick M, Siewe RM, Krämer R (1995) J Bacteriol 177:4690
146. Jebbar M, Talibart R, Gloux K, Bernard T, Blanco C (1992) J Bacteriol 174:5027
147. Fougère F, Le Rudulier D (1990) J Gen Microbiol 136:2503
148. Jung H, Jung K, Kleber HP (1990) J Basic Microbiol 30:409
149. Talibart R, Jebbar M, Gouesbet G, Himdi Kabbab S, Wriblewski H, Blanco C, Bernard T (1994) J Bacteriol 176:5210
150. Chambers ST, Kunin CM, Miller D, Hamada A (1987) J Bacteriol 169:4845
151. Milner JL, Grothe S, Wood JM (1988) J Biol Chem 263:14900
152. Stirling DA, Hulton CSJ, Waddell L, Park SF, Stewart GSAB, Booth IR, Higgins CF (1989) Mol Microbiol 3:1025
153. Lucht JM, Bremer E (1994) FEMS Microbiol Rev 14:3
154. Reed RH, Stewart WDP (1988) The response of cyanobacteria to salt stress. In:Rogers LJ, Gallon JR (eds) Biochemistry of the algae and cyanobacteria. Proceedings of the Phytochemical Society of Europe, Clarendon Press, Oxford, p 217
155. Peters P, Tel Or E, Trüper HG (1992) J Gen Microbiol 138:1993
156. Bae JH, Miller KJ (1992) Appl Environ Microbiol 58:471
157. Boch J, Kempf B, Bremer E (1994) J Bacteriol 176:5364
158. Stimeling KW, Graham JE, Kaenjak A, Wilkinson BJ (1994) Microbiology 140:3139
159. Pourkomailian B, Booth IR (1994) Microbiology 140:3131
160. Le Rudulier D, Strøm AR, Dandekar AM, Smith LT, Valentine RC (1984) Science 224:1064
161. Strøm AR, Falkenberg P, Landfald B (1986) FEMS Microbiol Lett 39:79
162. Styrvold OB, Falkenberg P, Landfald B, Eshoo MW, Bjørnsen T, Strøm AR (1986) J Bacteriol 165:856
163. Landfald B, Strøm AR (1986) J Bacteriol 165:849

164. Falkenberg P, Strøm AR (1990) Biochim Biophys Acta 1034:253
165. Yancey PH (1994) Compatible and counteracting solutes. In: Strange K (ed) Cellular and molecular physiology of cell volume regulation. CRC Press, Boca Raton, p 81
166. Arakawa T, Timasheff SN (1983) Arch Biochem Biophys 224:169
167. Arakawa T, Timasheff SN (1985) Biophys J 47:411
168. Collins KD, Washabaugh MW (1985) Quat Rev Biophys 18:323
169. Wiggins PM (1990) Microbiol Rev 54:432
170. Siew DCW, Cooney RP, Taylor MJ, Wiggins PM (1994) J Raman Spectroscopy 25:727
171. Wiggins PM (1994) Curr Topics Electrochem 3:129
172. Wiggins PM (1995) Cell Biochem Function 13:165
173. Wiggins PM (1995) Langmuir 11:1984
174. Wiggins PM (1995) Prog Polym Sci 20:1121
175. Galinski EA, Stein M, Amendt B, Kinder M (1996) Comp Biochem Biophys 117A:357
176. Lee CWB, Waugh JS, Griffin RG (1986) Biochemistry 25:3737
177. Gekko K, Timasheff SN (1981) Biochemistry 20:4667
178. Pollard R, Wyn Jones RG (1979) Planta 144:291
179. Muramatsu N, Minton AP (1988) Proc Natl Acad Sci USA 85:2984
180. Csonka LN, Epstein W (1996) Osmoregulation In: Curtis R III, Ingraham JL, Lin EEC, Low KB, Magasanik B, Reznikoff WS, Riley M, Schaechter M, Umbarger HE (eds) *Escherichia coli* and *Salmonella typhimurium*:cellular and molecular biology. Blackwell Science, Oxford, p 77
181. Nash D, Paleg LG, Wiskich JT (1982) Austral J Plant Physiol 9:47
182. Jolivet Y, Larher F, Hamelin J (1982) Plant Sci Lett 25:193
183. Pavlicek KA, Yopp JH (1982) Plant Physiol (Suppl) 69:58
184. Hurst A, El Banna AA, Hartwig J (1984) Can J Microbiol 30:1105
185. Manetas Y, Petropoulou Y, Karabourn[i]otis G (1986) Plant Cell Environ 9:145
186. Warr SRC, Reed RH, Stewart WDP (1984) J Gen Microbiol 130:2169
187. Back JF, Oakenfull D, Smith MB (1979) Biochem 18:5191
188. Paleg LG, Douglas TJ, v. Daal A, Keech DB (1981) Aust J Plant Physiol 8:107
189. Laurie S, Stewart GR (1990) J Exp Botany 41:1415
190. Lippert K, Galinski EA (1992) Appl Microbiol Biotechnol 37:61
191. Santoro MM, Liu Y, Khan SMA, Hou L, Bolen DW (1992) Biochem 31:5278
192. Gekko K (1982) J Biochem 91:1197
193. Fujita Y, Iwasa Y, Noda Y (1982) Bull Chem Soc (Japan) 55:1896
194. Krall JP, Edwards GE, Andreo CS (1989) Plant Physiol 89:280
195. van den Berg L, Rose D (1959) Arch Biochem Biophys 81:319; 84:305
196. Carpenter JF, Crowe JH (1988) Cryobiol 25:244
197. Carpenter JF, Hand SC, Crowe LM, Crowe JH (1986) Arch Biochem Biophys 250:505
198. Carpenter JF, Crowe LM, Crowe JH (1987) Biochim Biophys Acta 923:109
199. Carpenter JF, Martin B, Crowe LM, Crowe JH (1987) Cryobiol 24:455
200. Carpenter JF, Crowe JH (1988) Cryobiol 25:459
201. Crowe JH, Carpenter JF, Crowe LM, Anchordoguy TJ (1990) Cryobiol 27:219
202. Crowe JH, Crowe LM (1982) Cryobiol 19:317
203. Crowe JH, Crowe LM, Chapman D (1984) Science 223:701 703
204. Leslie SB, Israeli E, Lighthart B, Crowe JH, Crowe LM (1995) Appl Environ Microbiol 61:3592
205. Luard EJ (1983) Arch Microbiol 134:233
206. Shomer Ilan A, Waisel Y (1986) Physiol Plant 67:408
207. Yancey PH, Somero GN (1979) Biochem J 183:317

Received July 1997

Thermophilic Degradation of Environmental Pollutants

R. Müller[1]* · G. Antranikian[2] · S. Maloney[3] · R. Sharp[3]

[1] Arbeitsbereich Biotechnologie II, Technische Biochemie, Technische Universität Hamburg-Harburg, Denickestrasse 15, D-21071 Hamburg, Germany
E-mail: ru.mueller@tu-harburg.d400.de
[2] Arbeitsbereich Biotechnologie I, Technische Mikrobiologie, Technische Universität Hamburg-Harburg, Denickestrasse 15, D-21071 Hamburg, Germany
[3] CAMR (Centre for Applied Microbiology and Research) Porton Down, Salisbury, Wiltshire SP4 0JG, UK

This review summarizes present knowledge on the degradation of environmental pollutants at elevated temperatures. After a brief introduction to the relevance of this new emerging field of research, the current state of knowledge on the thermophilic conversion of alkanes, phenol and cresols, benzene and toluene, polycyclic aromatic hydrocarbons, pyrethroids, chlorinated hydrocarbons and nitro compounds is presented together with the information available on the degradation pathways and the enzymes involved.

Keywords: Thermophilic biodegradation, phenol, cresols, benzene, toluene, polycyclic aromatic hydrocarbons, pyrethroids, chlorinated hydrocarbons.

* Corresponding author.

Advances in Biochemical Engineering /
Biotechnology, Vol. 61
Managing Editor: Th. Scheper
© Springer-Verlag Berlin Heidelberg 1998

1
Introduction

Extreme thermophilic and hyperthermophilic microorganisms are classified as
those which are adapted to grow optimally at temperatures ranging from 70 to
110 °C. In the last decade it has been possible to isolate a number of microorganisms which belong to Archaea and Bacteria. Most of these microorganisms
that thrive above the boiling point of water belong to Archaea; the organisms
with the highest growth temperatures (103–110 °C) are members of the genera
Pyrobaculum, Pyrodictium, Pyrococcus and *Methanopyrus* [1]. Within the
Bacteria, *Aquifex pyrophilus* and *Thermotoga maritima* exhibit the highest growth temperatures of 95 and 90 °C, respectively. So far 54 species of hyperthermophilic Bacteria and Archaea are known [1]. They consist of anaerobic and
aerobic chemolithoautotrophs and heterotrophs, the latter being able to utilize
various polymeric substrates such as starch, hemicellulose, proteins and peptides
[2–4]. Most of thermophilic microorganisms that are able to grow at temperatures between 55 and 70 °C belong to the genera *Bacillus, Thermus* and *Clostridium*. Unlike hyperthermophiles these microorganisms may be isolated from a
wide range of environments including soils, sewage, composted vegetation,
river, lake and sea water, sediments and geothermal and hydrothermal environments [5]. Due to the high biodiversity of microbes living at a wide range of
elevated temperatures they are expected to play an important role in environmental sciences and biotechnology. The potential of thermophiles to be used in
the biodegradation of xenobiotics has been recognized only recently. In this
review we briefly describe the already known thermophilic microorganisms that
are able to degrade environmental pollutants.

1.1
Why Thermophilic Degradation?

Increase of temperature has a significant influence on the bioavailability and
solubility of organic compounds. The elevation of temperature is accompanied
by a decrease in viscosity and an increase in diffusion coefficient of organic
compounds. Consequently higher reaction rates due to smaller boundary layers
are expected. Of special interest are reactions catalysing the conversion of the
less soluble hydrophobic substrates such as polyaromatics, aliphatic hydrocarbons and fats, and polymeric compounds such as starch, cellulose, hemicellulose and proteins [2–4]. At elevated temperatures the solubility of such compounds is dramatically increased, allowing efficient bioconversion reactions due
to high substrate concentrations. Figure 1 shows the effect of temperature on the
solubility of the polyaromatic hydrocarbon anthracene in water. The bioavailability of hardly biodegradable and insoluble environmental pollutants can be
improved dramatically, allowing efficient bioremediation. By performing biological processes at temperatures above 60 °C the risk of contamination is reduced
and controlled processes under more defined conditions can be performed.
    The solubility of gases is influenced by temperature and it was commonly
believed that at higher temperatures the oxygen transfer rate (OTR) will decrea-

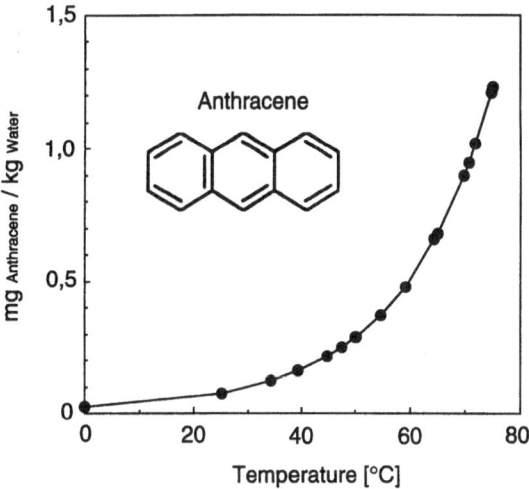

Fig. 1. The effect of temperature on the solubility of the polycyclic aromatic hydrocarbon anthracene in water

se due to lower oxygen solubility. Consequently, biological processes at elevated temperatures and under aerobic conditions would have been very limited. This, however, is not true since the OTR depends not only on the solubility of the gas but also on the volumetric mass transfer coefficient $k_L a$ as shown in the following equation [6]

$$OTR = k_L a (c^* - c_L)$$

where OTR = Oxygen transfer rate (mol m^{-3} h^{-1}), $k_L a$ = volumetric mass transfer coefficient for oxygen (h^{-1}), k_L = liquid film coefficient of absorption (m h^{-1}), a = volumetric area of the gas/liquid interface (m^2 m^{-3}), c^* = concentration of oxygen in the liquid, in equilibrium with the concentration in the gas phase (mol m^{-3}), and c_L dissolved oxygen concentration in the liquid phase (mol m^{-3}).

The diffusion coefficient for gases in liquid increases with temperature [7], and therefore, the mass transfer resistance due to the boundary layer of the liquid is smaller at higher temperature. The $k_L a$ values were determined as a function of temperature applying the dynamic pressure method with a downward pressure change of 0.2 bar [6, 8]. Figure 2 shows that the maximal OTR for air, calculated for dissolved oxygen concentration in the medium equal to zero, is slightly higher at 90 than at 30 °C. This clearly shows that supply of oxygen to the medium is the same or even better at elevated temperatures. According to these results it can be concluded that in principle biological processes for the degradation of hardly soluble xenobiotics is feasible at high temperatures and under aerobic conditions. Most of the bacteria with the ability to degrade xenobiotics under aerobic conditions at elevated temperatures that have been isolated so far belong to the thermophilic bacilli.

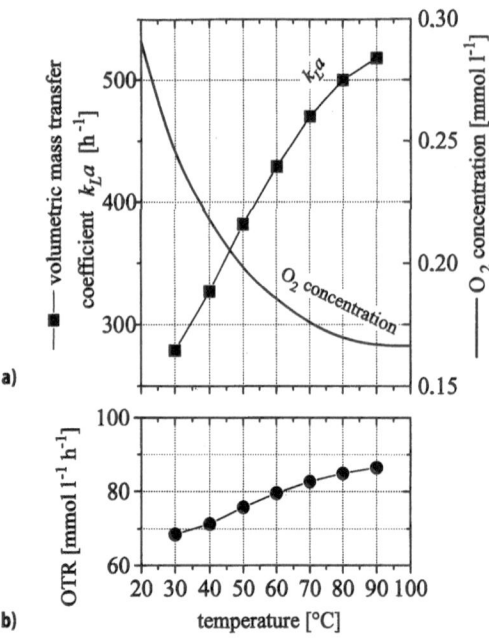

Fig. 2. a Solubility of oxygen in water as a function of the temperature. The mass transfer co-efficient k_La was measured in a 2-l Bioreactor at different temperatures. The aeration was performed with air at a rate of 1 vvm and a stirrer speed of 1500 rpm. **b** The resulting oxygen transfer rate (OTR) is shown as a function of temperature

1.2
Systematics of Thermophilic Bacilli

Thermophilic bacilli have been isolated from a wide range of environments including soil , sewage, composted vegetation, river, lake and sea water, sediments and many geothermal and hydrothermal environments [5]. They comprise a diverse group of species with a% G + C spanning 36–58% [9]. Most species are neutrophilic heterotrophs growing within the temperature range 30–78 °C. Some, including *B. acidocaldarius,* isolated from acid hot springs or geothermally heated soils and burning coal tips are able to grow at pH 2.0. *B. schlegelii* grows either chemolithotrophically using H_2, CO_2 or reduced sulphur compounds as electron donors and CO_2 as the carbon source, or chemoorganoheterotrophically using phenol, *i*-propanol and amino acids or organic acids as carbon source. Many species including *B. thermodenitrificans, B. flavothermus, B. thermocatenulatus* and some strains of *B. stearothermophilus* are able to grow in the absence of oxygen using nitrate as a terminal electron acceptor [5]. Other species described by Dijkhuizen et al. [10] and Al-Awadhi et al. [11] are able to utilise methanol, ethanol, propanol, and butanol as carbon sources.

With the ability to grow in such a wide diversity of environments and with a wide range of metabolic capabilities, thermophilic bacilli have been isolated in the presence of, being able to degrade, many environmental pollutants. Their important role in the breakdown of complex aromatic plant constituents in composting processes gives an indication of their degradative capabilities which can be harnessed for the breakdown of environmental pollutants with similar chemical structures.

2
Thermophilic Degradation of Pollutants

2.1
Alkanes

Alkanes are the major constituents of gasoline and other mineral oil products and are present in many polluted sites. They have a low toxicity and the solubility in water decreases with chain length. Under mesophilic conditions alkanes are biodegradable, although the low solubility is usually the limiting step in the biodegradation. Especially for alkanes of longer chain length, bioavailability due to low solubility becomes a problem. Therefore the degradation of alkanes at higher temperatures has been studied quite extensively. In 1967 Klug and Markovetz [12] described the isolation of a thermophilic bacterium with n-tetradecane as carbon source. In 1975 Seki examined the degradation of hydrocarbons in a natural pool of a hot spring [13]. He reported that hexadecane was utilised in this pond at 60 °C and the degradation rate was equivalent to that observed in natural waters at mesophilic temperatures.

In 1984 Zarilla and Perry [14] described the isolation of six strains that grew only on n-alkanes from 13 to 20 carbon atoms in length. Original isolation was carried out with n-heptadecane as substrate at 60 °C. These strains were Gram negative and did not form endospores. Due to their unique properties, especially the limited use of only alkanes as carbon sources, these strains were placed in a new genus named *Thermoleophilum.* The type strain is ATCC 35263. The properties of these strains were reviewed by Perry [15].

In 1987 the same authors [16] described 10 strains, which all grew on n-alkanes with 13–20 carbons. However, in contrast to *Thermoleophilum,* these strains formed endospores and were therefore named *Bacillus thermoleovorans.* In contrast to the *Thermoleophilum* strains these strains grew in addition to alkanes on various sugars and complex substrates such as casein, tryptone or yeast extract, but no growth was observed with 1-alkenes with 7–18 carbons or on 2-alkanones 5–12 carbons in length. Cycloalkanes and alkanes with less than 13 carbon atoms also did not support growth. Two of the strains were shown to produce bacteriocins [17]. These substances inhibited all but the producing strains from growth. These bacteriocins, termed thermoleovorins, were thermostable proteins with M_r of 42000 and 36000.

In recent years Sorkhoh et al. [18] and Nazina et al. [19] isolated thermophilic bacilli from oil fields in West Siberia and in the Kuwaiti desert. These strains grew on weathered crude oil. A closer examination of the substrate spectrum showed that these strains also degraded the n-alkanes with 14–17 carbon atoms preferentially. The aromatic compounds present in the crude oil were not used.

These strains also formed endospores and were classified as *Bacillus stearother-mophilus* and *Bacillus thermoleovorans*, respectively. In our laboratory we isolated thermophilic bacteria which were able to grow on much longer alkanes a feature not described for other strains [20]. The *n*-alkanes with 32 and 40 carbon atoms were good substrates for these bacteria. These substances are solid at room temperature. At 60 °C they are liquid and can therefore be much better distributed in the medium, thereby facilitating biodegradation.

2.2
Aromatic Compounds

2.2.1
Phenol and Cresols

Phenols are hazardous pollutants produced in oil refineries, petrochemical plants and various other industries. Due to their high toxicity to mesophilic microorganisms they are a source of constant concern in waste waters. Despite the fact that phenols can be easily degraded by mesophilic microorganisms, these organisms can not tolerate high concentrations of phenols. Concentrations as low as 0.25 mmol l^{-1} have been reported to inhibit bacterial growth. The pathways and the enzymes involved in phenol and cresol degradation by mesophilic microorganisms are well studied.

The utilisation of phenol and cresols by thermophilic microorganisms has also been known for some time. In 1975 Buswell and Twomey [21] described a *Bacillus stearothermophilus* that degraded phenol as well as all three isomers of cresol at 55 °C. Other aromatic compounds were not degraded. The degradation of all four substrates was induced simultaneously, indicating that all four compounds are degraded by the same enzymes. Interestingly the organism had a low tolerance towards phenol. Concentrations as low as 0.05 % were inhibitory. The phenols were all oxidised to the corresponding catechols, which were then cleaved in *meta* position. The resulting 2-hydroxymuconic acid semialdehyde was then either oxidised to 2-hydroxymuconic acid in the presence of NADH or, without NADH, formic acid was removed to form 2-hydroxypentadienoic acid [22, 23]. These pathways are identical to those observed in mesophilic microorganisms. In a survey of several thermophilic bacilli, Adams and Ribbons [24] found that five out of ten strains were able to use either benzoate or phenol as carbon source. None of these strains grew on any of the isomers of cresol or toluate.

In 1989 Gurujeyalakshimi and Oriel [25] described a *Bacillus stearothermophilus* that degraded phenol in concentrations up to 15 mmol l^{-1} at 55 °C. In this strain the enzymes for phenol degradation were encoded on a plasmid. The genes for the phenol hydroxylase as well as for the catechol 2,3-dioxygenase have been cloned. The growth kinetics of this strain were investigated intensively [26]. The discovery that the *meta* cleaving enzyme of this strain could be inhibited by tetracycline led to the suggestion of a thermophilic process for the biological production of catechol from phenol by the use of *Bacillus stearother-mophilus* BR219 in the presence of tetracycline [27]. Yanase et al. [28] isolated two strains from a brine sample from a submarine gas field, which degraded a variety of phenolic compounds. In addition to phenol several fluoro-, chloro-, methyl-, ethyl-, dimethyl- and trimethylphenols were converted by these strains. Due to the source of the strains it is not surprising that they required sea water for optimal activity.

We recently isolated a thermophilic bacillus that was able to degrade phenol as well as all three isomers of cresol at concentrations up to 10 mmol l^{-1}. This strain was able to grow up to 70 °C in mineral salts media containing the phenols as sole carbon source [29].

2.2.2
Benzene and Toluene

Although non-hydroxylated aromatic compounds are major constituents in many mineral oils, little information about their thermophilic degradation is available. Under mesophilic conditions these compounds are readily degraded. Many microorganisms have been isolated, the degradation pathways have been elucidated and the enzymes involved and the corresponding genes have been studied in detail [30]. One report describes the thermophilic degradation of benzene and toluene [31]. A *Bacillus stearothermophilus* isolated from the contaminated soil near a leaking petroleum pipeline was able to grow on benzene or toluene fed through the vapour phase. For the degradation of these compounds the addition of complex supplements such as yeast extract was necessary. Attempts to grow the strain on benzene without these additions failed. Which compounds of the yeast extract were necessary for growth is not known. For growth on phenol these compounds were not required. Recently Cheng and Taylor [32] described the cometabolic degradation of benzene, toluene and xylene by two *Thermus* spp. at 60 and 70 °C. However, in this study the concentrations of the substrates were very low, usually below 0.1 mmol l^{-1} and abiotic losses were rather high, often more than 50%. The degradation of the remaining portion varied between 10 and 40%. In experiments with radioactively labelled substrates, a slight increase in the water soluble fraction and traces of $^{14}CO_2$ were detected. Other metabolites were not detected. The cells were not able to grow during the conversion of the substrate. It remains to be seen whether these results can be confirmed by the identification of metabolites or by the detection of the degrading enzymes in *Thermus*.

2.2.3
Polycyclic Aromatic Hydrocarbons

Polycyclic aromatic hydrocarbons (PAH) are frequently found in soil contaminations derived from gas works. They are also present in many mineral oils and they are continuously formed in any combustion process. Today these compounds are found ubiquitously in the environment. Unfortunately many polycyclic aromatic compounds are carcinogenic and are therefore of great concern. The degradation of naphthalene, the simplest PAH, comprising only two rings, is well understood under mesophilic conditions [30]. The metabolites as well as

the enzymes involved in degradation and their genes are well known. On the degradation of the three- and four-ring PAH a lot of work has been done under mesophilic conditions. For the five- and six-ring PAH little information is available. This is due to the fact that these compounds with more rings are much less soluble and therefore are not available to microorganisms. Since it is known that their solubility greatly increases at higher temperatures, they should be ideal candidates for thermophilic degradation. However, so far no information is available on the thermophilic degradation of PAH.

In our laboratory we have isolated a bacterium which can grow on naphthalene as sole carbon source [20]. Initial studies on the metabolism indicate that metabolites are formed under thermophilic conditions which differ significantly from those formed under mesophilic conditions [33]. We also found strains that converted phenanthrene and anthracene under thermophilic conditions, but a closer examination of the degradation pathways has still to be done.

2.2.4
Pyrethroids

The photostable synthetic pyrethroids are an economically important group of insecticides [34]. They are widely used to control agricultural pests, and find increasing usage for the control of arthropods of medical, veterinary and industrial importance [34]. Pyrethroids have been developed to replace more toxic and environmentally persistent organochlorine, organophosphorus and methylcarbamate insecticides [34]. Despite some toxic effects on certain non-target organisms [35, 36], there is nothing to suggest that the importance and use of these insecticides is likely to decline in the foreseeable future. Previous studies have demonstrated microbially-mediated pyrethroid breakdown in soil and pure culture [37–39].

The contribution of thermophilic microorganisms, particularly anaerobes, to pollutant biodegradation has not been well investigated. Maloney et al. [40] reported the isolation and characterisation of a thermophilic Bacillus sp. able to transform a wide-range of pyrethroid insecticides to form non-insecticidal products (Fig. 3). More recently the microbially-mediated detoxification of five synthetic pyrethroid insecticides was demonstrated by a mixed culture growing anaerobically at 75 °C [41]. The mixed culture consisted of a facultative anaerobe and two obligate anaerobes including a methanogen. This culture was derived from terrestrial sediments from hot-springs in New Zealand.

Microorganisms that can operate at temperatures between 50 and 70 °C, and in some cases higher, would be of great benefit in developing end of pipe treatment systems for effluents which have not been considered to be biologically treatable [42]. The pyrethroid insecticides are often used as mothproofing agents and are found in effluents arising from carpet manufacturing and wool processing [36]. These insecticidally-active effluents are toxic to fish and often discharged at high temperatures. In addition these pesticides are poorly water-soluble and subsequently their availability for microbial degradation is limited under mesophilic conditions [34]. Increasing operating temperature enhances solubility, which may in turn lead to increased rates of biodegradation [40, 41].

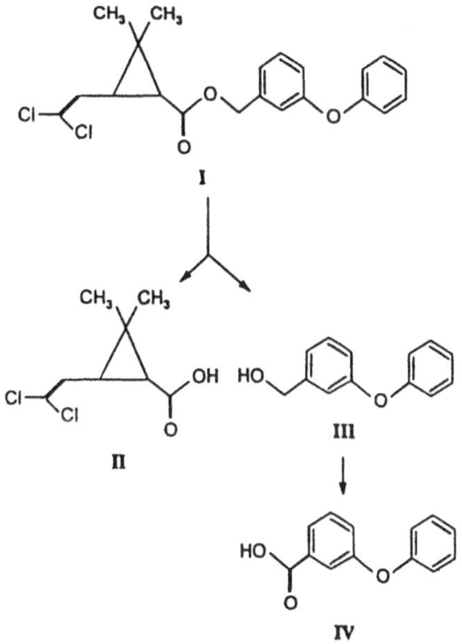

Fig. 3. Biotransformation of permethrin by thermophilic microorganisms at 75 °C. I Permethrin; II *trans* and *cis* isomers of 3-(2,2-dichlorovinyl)-2,2-dimethylcyclopropane carboxylic acid; III 3-phenoxybenzyl alcohol; IV 3-phenoxybenzoic acid

Enzymes from thermophiles are usually not only stable to heat, but also to organic solvents and detergents [43]. Therefore these microorganisms may prove more robust than mesophilic isolates when used in situ for the biological treatment of industrial effluents.

2.3
Chlorinated Hydrocarbons

Chlorinated hydrocarbons are widely used as solvents, insulating liquids, wood preservatives, herbicides and fungicides. Due to the stability of the carbon-chlorine bond, many of these substances are hardly biodegradable under mesophilic conditions. The highly chlorinated compounds pose special problems since the solubility of chlorinated hydrocarbons decreases with increasing chlorine content. Therefore these compounds should represent ideal targets for thermophilic degradation, where solubility is enhanced by the elevated temperature.

2.3.1
Chlorobenzoates

Chlorobenzoates are a by-product in the bacterial metabolism of polychlorinated-biphenyls [44] and herbicides [45].

Considerable research has been carried out on chlorobenzoate degradation under both aerobic and anaerobic conditions [46–50] in order to predict the fate of these chemicals in the environment and to develop techniques for the bioremediation of industrial wastes in polluted areas. A diverse range of sediment sources has been used in anaerobic studies, where microbes from different geographical locations have shown potential for chloroaromatic degradation [45, 51–53]. Reductive dehalogenation often represents the first step under anaerobic conditions and the lag-phase before reductive dehalogenation occurs is often quite long [54, 55].

The contribution of thermophilic microorganisms, particularly anaerobes, to reductive dehalogenation has not been well investigated. The microbial dehalogenation of 3-chlorobenzoate by mixed cultures (derived from both geothermal and non-geothermal sources) growing anaerobically at 75 °C is reported by Maloney et al. [56]. The dechlorination potential of the microbial population appeared not to be limited by its prior exposure to a particular xenobiotic or its source temperature, or type of environment. However, prior exposure to a xenobiotic may be important for accelerating dechlorination of these compounds by adaptation of the population. The consortium derived from the geothermal sample was maintained in an anaerobic continuous culture "bubble column" at 75 °C, to monitor changes in microbial composition and ability to dehalogenate other chlorinated organics at elevated temperatures. The lag period before dehalogenation of 3-chlorobenzoate in this study appeared shorter than that found in other studies [54, 55, 57]. It is possible that the thermophilic conditions may have affected the rate of dehalogenation, i.e. the reaction rate may be higher at elevated temperatures [58].

2.3.2
Other Chlorinated Hydrocarbons

Larsen et al. [59] demonstrated that different thermophilic ecosystems possess the ability for reductive dechlorination of pentachlorophenol. Briglia et al. [60] demonstrated the reductive dechlorination of hexachlorobenzene in river sediments and Michel et al. [61] showed that 2,4-dichlorophenoxyacetic acid can be degraded at 60 °C during composting of yard trimmings. In all cases the bacteria responsible for the degradation were not isolated. Therefore information as to which bacteria are responsible for the dehalogenation as well as information on the degradation pathways and the enzymes involved are not available.

Jablonski and Ferry [62] showed, that the thermophilic reduced CO-dehydrogenase complex from *Methanosarcina thermophila* reductively dechlorinated trichloroethylene via dichloroethylene and vinyl chloride to ethene.

We have demonstrated that thermophilic phenol-degrading microorganisms are capable of transforming chlorophenols cometabolically to the corresponding catechols [63] (Fig. 4) . These are not converted further by the ring cleaving enzymes, although at high temperatures they decompose spontaneously and do not poison the producing microorganisms as under mesophilic conditions (Fig. 5).

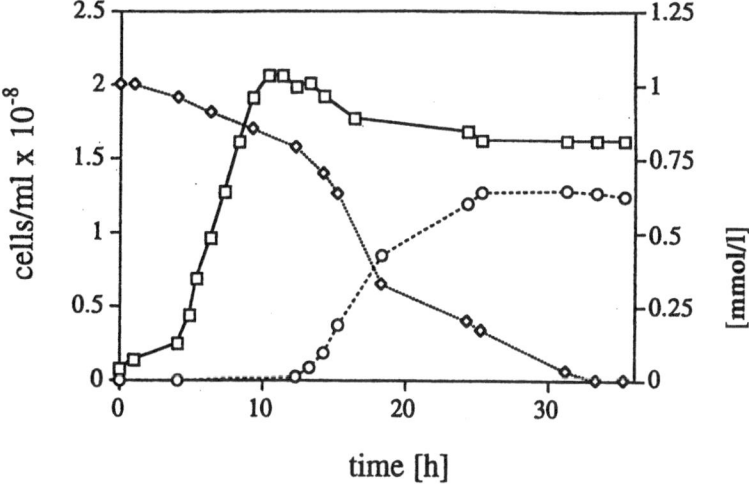

Fig. 4. Transformation of 1 mmol l⁻¹ 2-chlorophenol at 60 °C by *Bacillus* sp. A2—□—cells, —◇—2-chlorophenol, —○—3-chlorocatechol

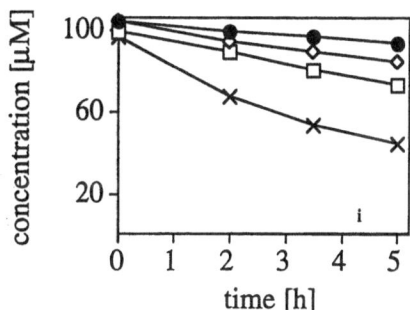

Fig. 5. Decomposition of 3-chlorocatechol at different temperatures —●— 40 °C, —◇— 50 °C, —□— 60 °C, —×— 70 °C

All the above-mentioned examples show that the potential for the thermophilic degradation of chlorinated compounds exists in nature. However, very little is known about the degradation mechanisms and the enzymes involved.

2.4
Nitro Compounds

Nitrated aromatic compounds are used as intermediates in dye and pesticide production. However, the biggest problems arise from their use as explosives. In particular, 2,4,6-trinitrotoluene (TNT) can be found in almost any site used for practice by an army.

Kaplan and Kaplan [64] demonstrated with ¹⁴C-labelled TNT that this compound is transformed in a thermophilic compost system. The reduction products isolated were similar to those found under mesophilic conditions. However a large portion of the substrate was incorporated into the humus fraction, whose structure could not be determined.

3
Pathways for the Degradation of Environmental Pollutants

So far only very little information is available on the pathways involved in the thermophilic degradation of the various compounds. In those cases where metabolites have been identified, the pathways seem to be identical to those described for mesophilic organisms. For example, in the degradation of phenol and cresol, the first steps, the hydroxylation and the ring cleavage in *meta* position, are identical to those under mesophilic conditions.

In other cases it seems reasonable to assume that the pathways are also identical to those under mesophilic conditions. For example, in the degradation of the alkanes no metabolites have been detected, but it may be assumed that the same pathway via monooxygenation of the terminal methyl group and further oxidation via the aldehyde to the fatty acid as it occurs under mesophilic conditions is employed. However a proof for this assumption is still missing.

In the case of naphthalene we found metabolites also isolated from mesophilic cultures, but in higher concentrations and we found metabolites not detected at lower temperatures [33]. Whether these new metabolites are due to chemical reactions occurring at higher temperatures or whether they are products of enzymatic reactions remains to be seen.

For many compounds nothing is known about their thermophilic degradation pathways, as in the case of PAH or 2,4-D.

4
Enzymes Involved in Thermophilic Degradation of Pollutants

The enzymes involved in the thermophilic degradation of pollutants should be interesting objects for research. They catalyse interesting reactions like oxidations of aromatic rings or dehalogenations and they should be active at elevated temperatures. This feature should make them interesting candidates for biotechnological applications. Surprisingly, practically no information on these enzymes is available. Only the first two enzymes in the degradation of phenol by *Bacillus stearothermophilus* have been detected and partially purified [25]. From *Bacillus stearothermophilus* BR219 and FDTP-3 the genes for these two enzymes have been cloned and sequenced. While the sequences from strain FDTP-3 [65] showed little homology to the corresponding genes from mesophilic microorganisms, the gene *phe*A from strain BR219 showed strong similarity to flavin hydroxylases from *Rhodococcus* and *Streptomyces* species [66].

We have cloned the phenol hydroxylase genes from two recently isolated *Bacillus* species. Again the sequences showed no significant homology to other phenol hydroxylases from mesophilic origin, but the sequences were also completely different from the two previously described thermophilic phenol hydroxylases. The highest degree of similarity was found with 4-hydroxyphenyl acetate 3-hydroxylase from *E. coli*. This enzyme requires for activity a small coupling protein. Interestingly we discovered that the phenol hydroxylase from our bacilli also requires a small protein for activity. Therefore the phenol hydroxylase of our thermophilic bacilli seem to be very different from all other phenol hydroxylases described so far.

We are sure that the characterisation of other enzymes from thermophilic microorganisms will lead to similar new discoveries.

5
Summary and Outlook

This paper illustrates that thermophilic microorganisms do possess a substantial potential for the degradation of environmental pollutants. The degradation of all major classes of environmental pollutants has been demonstrated. The organisms isolated in these studies may help to solve some of our environmental problems, but may also be interesting for new biotechnological applications.

In contrast, the knowledge on the thermophilic degradation pathways is still very limited and the properties of the enzymes involved are virtually unknown. Here is still an interesting area for future research, which needs to be exploited. New degradation pathways with new reactions catalysed by novel enzymes will be discovered, which will lead to novel biotechnological processes.

Acknowledgment. The authors are thankful to the Deutsche Forschungsgemeinschaft and Fonds der chemischen Industrie for financial support.

6
References

1. Stetter KO (1996) FEMS Microbiol Rev 18:149–158
2. Leuschner K, Antranikian G (1995) World J Microbiol Biotechnol 11:95–114
3. Sunna A, Moracci M, Rossi M, Antranikian G (1997) Extremophiles 1:2–13
4. Sunna A, Antranikian G (1997) Crit Rev Biotechnol 17:39–67
5. Sharp RJ, Riley PPW, White D (1992) In: Kristjansson JK (ed) Thermophilic bacteria, pp 19–50
6. Krahe M, Antranikian G, Märkl H (1996) FEMS Microbiol Rev 18:271–285
7. Perry RH, Green DW (1984) Perry's Chemical Engineers' Handbook, 6th ed. McGraw Hill, New York, pp 3.101–3.103
8. Linek V, Moucha T, Dousova M, Sinkule J (1994) J Biotechnol Bioeng 43:477–482
9. White D, Sharp RJ, Priest FG (1993) Antonie van Leeuwenhoek 64:357–386
10. Dijkhuizen L, Arfman N, Attwood MM, Brooke AG, Harder W, Watling EM (1988) FEMS Microbiol Lett 52:209
11. Al-Awadhi N, Egli T, Hamer G, Wehrli E (1989) Syst Appl Microbiol 11:207
12. Klug MJ, Markovetz AJ (1967) Nature 215:1082–1083
13. Seki H (1975) La Mer 13:53–57
14. Zarilla KA, Perry JJ (1984) Arch Microbiol 137:286–290
15. Perry JJ (1985) Adv Aquatic Microbiol 3:109–139
16. Zarilla KA, Perry JJ (1987) Syst Appl Microbiol 9:258–264
17. Novotny JF, Perry JJ (1992) Appl Environ Microbiol 58:2393–2396

18. Sorkhoh NA, Ibrahim AS, Ghannoum MA, Radwan SS (1993) Appl Microbiol Biotechnol 39:123-126
19. Nazina TN, Ivanova AE, Mityushina LL, Belyaev SS (1993) Microbiol 62:359-365
20. Hebenbrock S, Feitkenhauer H, Müller B, Märkl H, Antranikian G (1996) Biospektrum Sonderausgabe 1996:108
21. Buswell JA, Twomey DG (1975) J Gen Microbiol 87:377-379
22. Buswell JA (1974) Biochem and Biophys Res Commun 60:934-941
23. Buswell JA (1975) J Bacteriol 124:1077-1083
24. Adams D, Ribbons DW (1988) Appl Biochem Biotechnol 17:231-244
25. Gurujeyalakshimi G, Oriel P (1989) Appl Environ Microbiol 55:500-502
26. Worden R-M, Submaranian R, Bly MJ, Winter S, Aronson CL (1991) Appl Biochem Biotechnol 28/29:267-275
27. Gurujeyalakshimi G, Oriel P (1989) Biotechnol Lett 2:689-694
28. Yanase H, Zuzan K, Kita K, Sogabe S, Kato N (1992) J Ferment Bioeng 74:297-300
29. Mutzel A, Reinscheid U, Antranikian G, Müller R (1996) Appl Microbiol Biotechnol 46:593-596
30. Müller R, Lingens F (1983) In: Sund H, Ullrich V (eds) Biological oxidations. Springer, Berlin Heidelberg New York, pp 278-287
31. Natarajan MR, Lu Z, Oriel P (1994) Biodegradations 5:77-82
32. Cheng C-I, Taylor RT (1995) Biotechnol Bioeng 48:614-X624
33. Müller B, Hebenbrock S, Feitkenhauer H, Steinhart H, Antranikian G, Märkl H (1996) In: 1st international congress on extremophiles, Estoril, Portugal, p 165
34. Leahey JP (1985) The Pyrethroid Insecticides, Taylor and Francis, London
35. Jolly AL, Graves JB, Avault JW, Koonce KL (1977/8) Louisiana Agriculture 21:3-4
36. Zabel TF, Seager J, Oakley SD (1988) Water Research Council, Environmental Strategy, Standards and Legislation Unit, TR 261, Medmenham
37. Kaufman DD, Haynes SC, Jordan EG, Kayser AJ (1977) In: Eliot M (ed) Synthetic Pyrethroids. American Chemical Society Series No. 42, San Francisco, CA
38. Maloney SE, Maule A, Smith ARW (1988) Appl Environ Microbiol 54:2874-2876
39. Maloney SE, Maule A, Smith ARW (1993) Appl Environ Microbiol 59:2007-2013
40. Maloney SE, Maule A, Smith ARW (1992) Arch Microbiol 158:282-286
41. Maloney SE, Marks TS, Sharp RJ (1997) J Chem Technol Biotechnol 68:357-360
42. Cowan DA (1992) Biochemical Society Symposium 58:149-169
43. Coolbear T, Daniel RM, Morgan HW (1992) Adv Biochem Eng Biol 45:57-98
44. Furukawa K, Tomizuka N, Kamibayashi A (1983) Appl Environ Microbiol 46:140-145
45. Häggblom MM (1992) FEMS Microbiol Rev 103:29-72
46. Suflita JM, Horowitz A, Shelton DR, Tiedje JM (1982) Science 218:1115-1116
47. Horowitz A, Suflita JM, Tiedje JM (1983) Appl Environ Microbiol 45:1459-1465
48. Marks TS, Smith ARW, Quirk AV (1984) Appl Environ Microbiol 48:1020-1025
49. DeWeerd KA, Mandelco L, Tanner RS, Woese CR, Suflita JM (1990) Arch Microbiol 154:23-30
50. Gerritse J, Woulde BJ vd, Gottshal JC (1992) FEMS Microbiol Lett 100:273-280
51. Shelton DR, Tiedje JM (1984) Appl Environ Microbiol 48:840-848
52. Sharak Genthner BR, Price IAW, Pritchard PH (1989) Appl Environ Microbiol 55:1466-1471
53. Kazumi J, Haeggblom MM, Young LY (1995) Appl Microbiol Biotechnol 43:929
54. Battersby NS, Wilson V (1989) Appl Environ Microbiol 55:433-439
55. Linkfield TG, Suflita JM, Tiedje JM (1989) Appl Environ Microbiol 55:2773-2778
56. Maloney SE, Marks TS, Sharp RJ (1997) Lett Appl Microbiol 24:441-444
57. Maloney SE, Laughton JC, Maule A, Sharp RJ (1996) Dept of the Environment Rep CWM/107/94L
58. Adams MWW, Perler FB, Kelly RM (1995) Bio/technology 13:662-668
59. Larsen S, Hendriksen HV, Ahring BK (1991) Appl Environ Microbiol 57:2085-2090
60. Briglia M, Schraa G, de Vos WM (1995) Biospektrum 3/95:71
61. Michel FC, Reddy CA, Forney LJ (1995) Appl Environ Microbiol 61:2566-2571

62. Jablonski PE, Ferry JG (1992) FEMS Microbiol Lett 96:55–60
63. Reinscheid UM, Bauer MP, Müller R (1996) Biodegradation 7:455–461
64. Kaplan DL, Kaplan AM (1982) Appl Environ Microbiol 44:757–760
65. Dong F, Wang L, Wang C, Cheng J, He Z, Sheng Z, Shen R (1992) Appl Environ Microbiol 58:2531–2535
66. Kim IC, Oriel PJ (1995) Appl Environ Microbiol 61:12512–1256

Received June 1997

Author Index Volume 1–61

Elling, L.: Glycobiotechnology: Enzymes for the Synthesis of Nucleotide Sugars. Vol. 58, p.89

Fiechter, A. see Ochsner, U. A.: Vol. 53, p. 89
Freitag, R., Hórvath, C.: Chromatography in the Downstream Processing of Biotechnological Products. Vol. 53, p. 17
Eriksson, K.-E. L. see Kuhad, R. C.: Vol. 57, p. 45
Eriksson, K.-E. L. see Dean, J. F. D.: Vol. 57, p. 1

Farrell, R. L., Hata, K., Wall, M. B.: Solving Pitch Problems in Pulp and Paper Processes. Vol. 57, p. 197

Galinski, E. A. see da Costa, M. S.: Vol. 61, p. 117
Gatfield, I.L.: Biotechnological Production of Flavour-Active Lactones. Vol. 55, p.221
Gerlach, S. R. see Schügerl, K.: Vol. 60, p. 195
Ghosh, A. C., Mathur, R. K., Dutta, N. N.: Extraction and Purification of Cephalosporin Antibiotics. Vol. 56, p. 111
Ghosh, P. see Singh, A.: Vol. 51, p. 47
Gomes, J., Menawat, A. S.: Fed-Batch Bioproduction of Spectinomycin. Vol. 59, p. 1
de Graaf, A.A. see Eggeling, L.: Vol. 54, p. 1
de Graaf, A.A. see Weuster-Botz, D.: Vol. 54, p. 75
de Graaf, A.A. see Wiechert, W.: Vol. 54, p. 109
Gros, J.-B. see Larroche, C.: Vol. 55, p. 179
Gros, J.-B. see Cornet, J. F.: Vol. 59, p. 153
Guenette M. see Tolan, J. S.: Vol. 57, p. 289
Gutman, A. L., Shapira, M.: Synthetic Applications of Enzymatic Reactions in Organic Solvents. Vol. 52, p. 87

Hall, D. O. see Markov, S. A.: Vol. 52, p. 59
Hasegawa, S., Shimizu, K.: Noninferior Periodic Operation of Bioreactor Systems. Vol. 51, p. 91
Hata, K. see Farrell, R. L.: Vol. 57, p. 197
Hembach, T. see Ochsner, U. A.: Vol. 53, p. 89
Hill, D. C., Wrigley, S. K., Nisbet, L. J.: Novel Screen Methodologies for Identification of New Microbial Metabolites with Pharmacological Activity. Vol. 59, p. 73
Hiroto, M. see Inada, Y.: Vol. 52, p. 129
Hórvath, C. see Freitag, R.: Vol. 53, p. 17
Hummel, W.: New Alcohol Dehydrogenases for the Synthesis of Chiral Compounds. Vol. 58, p.145

Inada, Y., Matsushima, A., Hiroto, M., Nishimura, H., Kodera, Y.: Chemical Modifications of Proteins with Polyethylen Glycols. Vol. 52, p. 129
Johnson, E. A., Schroeder, W. A.: Microbial Carotenoids. Vol. 53, p. 119
Joshi, J. B. see Elias, C. B.: Vol. 59, p. 47
Johnsurd, S. C.: Biotechnolgy for Solving Slime Problems in the Pulp and Paper Industry. Vol. 57, p. 311

Kataoka, M. see Shimizu, S.: Vol. 58, p. 45
Kawai, F.: Breakdown of Plastics and Polymers by Microorganisms. Vol. 52, p. 151
King, R.: Mathematical Modelling of the Morphology of Streptomyces Species. Vol. 60, p. 95
Kirk, T. K. see Akhtar, M.: Vol. 57, p. 159
Kobayashi, M. see Shimizu, S.: Vol. 58, p. 45
Kodera, F. see Inada, Y.: Vol. 52, p. 129
Krabben, P. Nielsen, J.: Modeling the Mycelium Morphology of Penicilium Species in Submerged Cultures. Vol. 60, p. 125
Krämer, R.: Analysis and Modeling of Substrate Uptake and Product Release by Procaryotic and Eucaryotik Cells. Vol. 54, p. 31

Subject Index

Springer
and the
environment

At Springer we firmly believe that an international science publisher has a special obligation to the environment, and our corporate policies consistently reflect this conviction.

We also expect our business partners – paper mills, printers, packaging manufacturers, etc. – to commit themselves to using materials and production processes that do not harm the environment. The paper in this book is made from low- or no-chlorine pulp and is acid free, in conformance with international standards for paper permanency.

 Springer